THE PHYSICS OF LASER FUSION

THE PHYSICS OF LASER FUSION

H. MOTZ

Clarendon Laboratory,
Oxford

1979

ACADEMIC PRESS
London · New York · San Francisco
A Subsidiary of Harcourt Brace Jovanovich, Publishers.

U.K. Edition published and distributed by
ACADEMIC PRESS INC. (LONDON) LTD.
24/28 Oval Road
London NW1

United States Edition published and distributed by
ACADEMIC PRESS INC.
111 Fifth Avenue
New York, New York 10003

British Library Cataloguing in Publication Data

Motz, Hans
    The physics of laser fusion.

    1. Laser fusion
    I. Title
    539.7'64   QC791.73   78-67904
    ISBN 0-12-509350-0

Printed in Gt. Britain by
Page Bros (Norwich) Ltd, Norwich

PREFACE

Are fusion reactions feasible? The answer to this question may turn out to be of great significance for the future of mankind. It will determine the options open for the task of providing energy to sustain civilized life on the planet. Existing science cannot give a definitive answer to this question. Fusion reactions sustain the sun and have done so for times of the order of 10^9 years. Hydrogen bombs are fusion devices liberating the energy within microseconds. A controlled thermonuclear device for the use of fusion energy must work on a time scale in between these extremes. It must be of a suitable size and deliver a suitable amount of power, say between 1 MW and 10^4 MW per plant with a capital expenditure comparable to or less than that needed by its competitors. Thus even if the question of whether energy output of a fusion reaction can be made to exceed energy input in a terrestrial device is answered affirmatively, a number of technological and economic questions remain to which physics has no ready answer.

Yet, the enthusiasm and faith of the scientific community has commanded support for a large research effort in the fusion field. Broadly there are two approaches: magnetic confinement of hot plasma and inertial confinement. Inertial confincment schemes aim at compression of deuterium–tritium fuel to very high densities, so that the reaction can proceed extremely quickly, within pico- or nanoseconds, as a microexplosion.

This compression may be achieved by a flux of energetic particles (electrons or ions) or by means of laser light. This book is mainly devoted to laser driven fusion. It covers a whole range of interesting problems of physics; they range from Plasma Physics to Fluid Dynamics and the properties of plasma at the high densities, temperatures and pressures encountered in stellar matter. It seemed desirable to introduce novices in this field to relevant physics topics. In view of their wide range, it is hoped that even the specialist might find some chapters useful. They are written in such a way that they can be read independently, particularly by readers acquainted with the rest of the material. The book is thus written for physicists or engineers of divers backgrounds and aims at a consistent, easily readable presentation. Derivations of results will at least be sketched, leaving out details if they can easily be found in existing texts, which need not be consulted in order to comprehend this book.

Questions of a more technological nature are not tackled. The emphasis is on the understanding of the basic physical processes. They are rather involved and in order to get realistic quantitative theoretical predictions it is necessary to resort to numerical computations. It is unfortunately rather difficult to gain an insight into what is going on without guidance by analytic solutions of simplified processes which intermingle in the real situation. In this book, analytically soluble problems which are relevant to laser fusion are presented and solved. The results of full computer computations are reported as much as is possible within our limited space, and the physics underlying the computational codes is explained.

Perhaps a few words on how to read the mathematics in this book will be helpful. When some intermediate steps are left out, they are either too trivial or too long-winded. It is best to concentrate on the statement that a given equation makes the relationship which it establishes. The aim is throughout to make clear what kind of reasoning leads to the result and to quote results when the algebra is rather involved.

February 1979   H. Motz
                                                                 Oxford

CONTENTS

ACKNOWLEDGMENTS

Many colleagues have helped me with the preparation of this book and I wish to thank them. Dr R. N. Franklin, who has read most of the manuscript offered valuable criticism, Dr J. E. Allen and Mr L. M. Wickens, Dr M. H. Key, Dr T. P. Hughes, Prof. E. A. Mishkin, Dr H. E. DeWitt, Prof. J. P. Hansen, Prof. N. H. March, Dr P. T. Rumsby, Dr J. Kilkenny, Dr D. B. Henderson, Dr S. B. Segall have contributed important material. I especially want to thank Prof. A. F. Gibson Director of the Rutherford Laboratory for allowing me to reproduce Figs 2 and 3 of Chapter 3 as well as other material from the reports of the laboratory and Dr H. G. Ahlstrom of the Livermore Laboratory, who communicated much of the material of Chapter 13. My thanks are also due to Mrs E. Rose, Librarian.

1

FUSION REACTIONS

In the early days of nuclear physics it was already known that energy may be released by reactions between light nuclei (fusion reactions) or by splitting a very heavy nucleus into two or more parts (fission reactions). In either case these energies are of the order of MeV per nucleon which is six orders of magnitude more than those available per atom of chemical reactions.

When the sum of the masses of light nuclei (e.g. protons and neutrons) exceeds the mass of the product nucleus by an amount δm, the energy W liberated by this fusion of nuclei is, by Einstein's relation, given by

$$W = \delta m . c^2$$

and it is released in the form of kinetic energy of the reaction products. For such a reaction to take place, the particles must come within range of the nuclear forces and this means that the Coulomb barrier has to be overcome by the kinetic energy available in the centre of mass system of the colliding particles. It was soon realized that bombardment of light-element targets with high-energy particle beams could not efficiently produce power, unless the energy, necessarily imparted to outer shell electrons in the collision process, was utilized. This means that the reacting particles must be confined at high density for a time sufficiently long for energy transfer to the nuclei to take place. This transfer is disorderly, i.e. the ions (or nuclei) are heated to a high temperature. At a given temperature, the reaction rate is proportional to the square of the density; the time during which confinement can be secured turns out to be limited to a small fraction of a second and, therefore, the density needed in order to achieve a useful power output is very high.

The temperature required for barrier penetration and the density required for a practical device will be determined below from data concerning reaction cross-sections. They represent conditions of matter known to exist in stars. The H-bomb first realized similar conditions on earth and the problem of its use for triggering a thermonuclear reaction was taken up and solved by E. Teller. The first release of man-made thermonuclear energy occurred in 1952 but the problem of how to control this release for the purpose of generating electric power is still with us. Early reviews by Post [1] and by Glasstone and Lovberg [2] may be consulted for an introduction to the main lines of research on controlled thermonuclear reactions which aim at a quasi-steady-state confinement by magnetic fields. The main difficulty of this approach is well known. It is necessary to produce a state of matter far from equilibrium where large pressure and temperature gradients cannot easily be maintained in the face of instabilities causing premature particle and energy loss. This book is concerned with the basic physics underlying another approach: the pulsed energy-release from a fuel pellet, with no external confinement during the very short time required for the fluid dynamic motion in the pellet during which it is compressed to a high density such that the reaction rate is high enough to lead to a useful output.

The problems associated with magnetic confinement thus cannot arise, but the method has problems of its own which will emerge in the further course of the exposition. But first we want to examine the conditions which must be realized to make this approach work.

To do this we must assemble information concerning the nuclear reactions involved. The basic reactions are

$$D + D \rightarrow {}^3He\,(0.82\ MeV) + n\,(2.45\ MeV) \tag{1}$$

$$D + D \rightarrow T\,(1.01\ MeV) + H\,(3.02\ MeV) \tag{2}$$

$$D + T \rightarrow {}^4He\,(3.5\ MeV) + n\,(14.1\ MeV) \tag{3}$$

$$D + {}^3He \rightarrow {}^4He\,(3.6\ MeV) + H\,(14.7\ MeV) \tag{4}$$

and the division of kinetic energy between the reaction products has been indicated. As will be seen below, the last reaction becomes interesting only at higher energies. There are other reactions involving lithium and boron isotopes which may be of interest at high temperatures, e.g.

$$B^{11} + H \rightarrow He + He + He$$

which involves only charged reaction products and does not lead to radioactive contamination, but from a practical point of view, only the DD and the DT reactions have been considered. We shall see that the DT reaction is much more favourable because it occurs at lower energies than the DD reaction. On the other hand, deuterium occurs in nature with an abundance of 1 in 6500 atoms while tritium which is radioactive with a half-life of 12.26 yr has to be manufactured. The neutrons from the thermonuclear reactions could be slowed down in a suitable moderator containing lithium-6 to produce tritium by the reaction

$$^6\text{Li} + {}_0n \rightarrow {}_2\text{He} + {}_1\text{T} + 4\cdot6 \text{ MeV} \tag{5}$$

in the fusion plant itself.

The reaction rates can be estimated by means of theoretical considerations due to Houtermans et al. [3] and Gamov [4]. The energy needed to surmount the Coulomb barrier is given by $Z_A Z_B e^2 / R_0$ in terms of the nuclear charges Z_A and Z_B of the reaction partners and the range R_0 of the nuclear forces between them. Taking this equal to a nuclear diameter 5.10^{-13} cm and with $e = 1\cdot6.10^{-19}$ C, this becomes $0\cdot28 Z_A Z_B$ MeV. Thus classical theory would predict a minimum energy of $0\cdot28$ MeV for hydrogen isotopes. Quantum theory abolishes the sharp limit and there is no threshold. The probability of barrier penetration as a function of energy E is given by

$$\sigma_{AB}(E) \cong \frac{A}{E} \exp(-2^{3/2}\pi^2 m^{1/2} Z_A Z_B e^2 / hE^{1/2}) \tag{6}$$

where m is the reduced mass $m = m_A m_B/(m_A + m_B)$, h is Planck's constant, and E is the total energy in the centre of the mass system of the nuclei A and B. The reaction rate in a hot gas with nuclei of type A and velocity $\mathbf{u}$ and of type B with velocity $\mathbf{v}$ is defined by the double integral over the velocity space spanned by $\mathbf{u}$ and $\mathbf{v}$

$$R_{AB} = \int\int f(\mathbf{u}) \, g(\mathbf{v}) \, \sigma(w) w \, d\mathbf{u} \, d\mathbf{v} \tag{7}$$

where $f(\mathbf{u})$ and $g(\mathbf{v})$ are the respective velocity distributions functions such that the particle numbers in the volume elements $d\mathbf{u}$ and $d\mathbf{v}$ of velocity per unit volume of ordinary space are given by

$$dn_A = f(\mathbf{u}) \, d\mathbf{u}$$
$$dn_B = g(\mathbf{v}) \, d\mathbf{v} \tag{8}$$

and where w is the relative velocity

$$w = |\mathbf{u} - \mathbf{v}|.$$

For Maxwellian velocity distributions it follows from (6) that in a system where the nuclei B are at rest and the nuclei A have velocity v and kinetic energy $E = m_A v^2/2$;

$$R_{AB} = \langle \sigma v \rangle = (8\pi n_A n_B/m_A^2)(m/2\pi kT)^{3/2} \int E\sigma_{AB}(E)$$

$$\times \exp(-mE/m_A kT)) \, dE. \tag{9}$$

The result (9) of folding the cross-section $\sigma(E)$ into the distribution function, "the velocity averaging" is, in the literature commonly called $\langle \sigma v \rangle$ or $\overline{\sigma v}$. $\overline{\sigma v}$.

The constant of formula (6) depends on R_0 and must thus be determined experimentally. The fusion cross-section may be written

$$\ln E\sigma_{AB} = A' - E^{-1/2} B', \tag{10}$$

the constants A' and B' have been determined from experiments, and the resulting reaction rates have been computed by several authors [5–8] who are not in complete agreement. The useful curves of Fig. 1 reproduced from

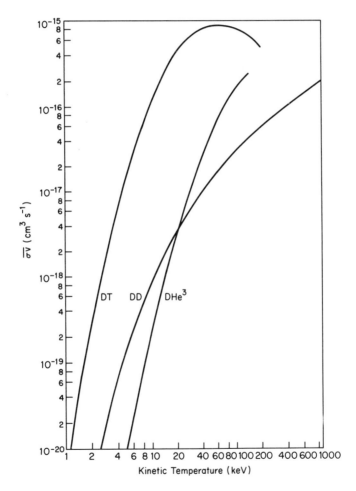

Fig. 1. Values of $\langle \sigma v \rangle$ based on Maxwellian distribution for DT, DD (total), and DHe³ reactions.

[2] show the essential features. The temperature is expressed in keV (1 keV → 8·6 . 10^7 K) which is usual in the literature on fusion. The DT rate in the range up to 60 keV is orders of magnitude higher than the DD rate and at 9 keV it is already up to 10% of the maximum rate.

Post remarks [1] that the "tail of the Maxwell distribution wags the dog." This is seen in Fig. 2 which shows the contributions of deuterons of various energies for the DD reaction (2) which is expressed as $\langle \sigma v \rangle = \int_0^\infty \phi(E)\, \mathrm{d}E$.

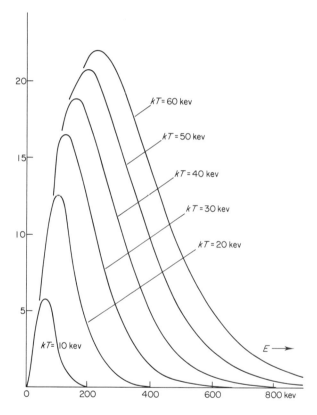

Fig. 2. Contribution to the reaction rate R_{DD} as a function of the deuteron energy E for kT = (left to right) 10, 20, 30, 50, 60 keV.

The DT rate behaves similarly. Figure 2 shows ϕ_{DD} for temperatures of 10–60 keV, in steps of 10 keV. It is seen that the reaction rate is in each case mainly due to particles belonging to the tail of the Maxwell distribution.

An approximation to the reaction rate can be obtained by means of a method due to Gamov. One can write, in first approximation

$$\int_0^\infty \phi(E)\, dE = \Delta E \phi(E_m), \tag{11}$$

because the integrand is of the form

$$\phi(E) = \exp\{-(mE/m_A kT) - B'E^{-1/2}\} \tag{12}$$

and has a sharp maximum as a function of E at $E = E_m = (B'm_A kT/2m)^{3/2}$. If ΔE is defined by

$$\phi(E + \Delta E/2) = \phi(E_m)/e, \tag{13}$$

ΔE is given by

$$\Delta E = (3^{2/3} B')^{1/2} \qquad (B'm_A kT/2m)^{5/6}. \tag{14}$$

In the results for the DD and DT reaction rates which have been tabulated below, the values for E_m, the energy of the maximum contribution to the reaction rate, have been entered, together with the computed reaction rates for various temperatures.

In Table I for the DD reaction, we have included the values of E_m and R_{DD} as computed from the Gamov approximation. It is seen that the Gamov approximation holds only for low temperatures.

TABLE I. R_{DD} in units of 10^{-18} cm^{-3} s^{-1} $(nD\ \text{cm}^3)^2$

kT (keV)	E_m (Gamov) (keV)	E_m (keV)	R_{DD} (Gamov)	R_{DD}	Percent Effective
10	59	50	0·60	0·98	11
20	94	104	2·38	3·40	16
30	124	124	4·45	6·75	24
40	150	134	6·46	10·11	34·2
50	174	185	8·33	13·87	29·6
60	196	225	10·03	17·96	29·7

TABLE II. R_{DT} in units of 10^{-18} $(nD nT\ \text{cm}^6)$ cm^{-3} s^{-1}

kT (keV)	E_m (keV)	R_{DT}	Percent effective
2	19·8	0·263	0·71
4	31·6	5·91	2·3
6	40·6	25·71	4·27
7	44·4	40·27	5·4
8	48	60·51	6·52
40	110	748	25

It is interesting to compute the mean energy $m_D \bar{u}^2/2$ of deuterons in the laboratory frame of reference which have relative energy $E = m_D W^2/2$ on collision. Statistical analysis gives the result

$$m_A \bar{u}^2/2 = \tfrac{1}{2} m_A W^2 (m/m_A)^2 + \tfrac{3}{2} kT(m/m_B) \qquad (15)$$

where m is the reduced mass.

Conversely, one can compute the mean relative energy on collision of deuterons with energy $m_D u^2/2$ in the laboratory system. The result is

$$m_A \overline{W}^2/2 = m_A u^2/2 + \tfrac{3}{2} kT(m_A/m_B). \qquad (16)$$

Using these results, Tables III and IV can be computed for DD and DT reactions

TABLE III. DD reactions.

kT (keV)	E_m (keV)	$mD\bar{u}^2/2$ (keV)	$mDu^2/2$ (keV)
10	59	22	44
20	94	30	64
30	124	54	79
40	150	68	90
50	174	81	99
60	196	94	106

TABLE IV. DT reactions

kT (keV)	E_m (keV)	$mD\bar{u}^2/2$ (keV)	$mDu^2/2$ (keV)
2	19·8	8·35	17·8
4	31·6	14·1	27·6
6	40·6	18·2	34·6
7	44·4	20·2	37·4
8	48·0	22·1	40
40	110	63·8	70

The energies $m_D u^2/2$ have been computed from equation (16), where $m_A \overline{W}^2/2$ has been identified with E_m. They are thus the energies of deuterons which have a mean relative energy equal to the relative energy giving the maximum contribution to the reaction rate.

In Tables I and II there are columns headed "Percent Effective". These are the percentages $\Delta n_A/n_A \cdot 100$ of deuterons with relative energy larger than E_m

computed from

$$\Delta n_A / n_A = n_B 4 / \sqrt{\pi} \int_Z^\infty z^2 \exp(-z^2)\,\mathrm{d}z \qquad (17)$$

where the lower limit of the integral is given by $Z = (mW^2/2)^{1/2}$ and m is again the reduced mass.

Looking at these tables, one notes that ions with fairly low energies in the laboratory system have on average a high relative energy. This may be relevant when one is considering the effect on the reaction rate of accelerated deuterons inserted into a discharge. Looking at Table II one notes that in the case of the DT reaction, the effective percentages are rather low, again showing that relatively few accelerated deuterons injected into a plasma may produce reaction rates comparable to the purely thermonuclear ones.

ENERGY LOSS AND RADIATION BY PARTICLES

Fusion devices operate at temperatures in the keV range. In this range matter is completely ionized, ions of low mass number are stripped of their orbital electrons; it consists of a gas of positively charged nuclei and an equivalent number of electrons which is referred to as a plasma. (More precisely an ionized gas is called a plasma if the Debye length (Chapter 4, equation (8)) is small compared to all other lengths of interest).

Hydrogen atoms are completely stripped and energy loss occurs in the form of *bremsstrahlung*, i.e. continuous radiation emitted by charged particles, mainly electrons, as a result of deflection by the Coulomb field of other charged particles. Ions of higher Z, if they are not completely stripped, are subjected to excitation processes contributing to radiation loss. These losses must be subtracted from the yield of the nuclear reactions and we shall compute the loss rate below.

The deflections suffered by protons and α-particles during their passage through the plasma are also of great interest. They limit their range in such a way that, at sufficiently high density, their energy may be deposited in the fusion pellet thus accelerating the burning process.

Rather than just quoting the relevant formulae we proceed to an intuitive derivation. The classical expression for the rate P_e at which energy is radiated by an accelerated electron is given by

$$P_e = \frac{2}{3} \frac{e^2}{c^3} a^2. \qquad (18)$$

The acceleration a due to the Coulomb force between charged particles separated by a distance b is given by

$$a = \frac{Ze^2}{b^2 m_e}. \tag{19}$$

We have put the mass equal to the mass of an electron which is considered to move past a stationary ion of charge Ze and the rate of energy loss is then given by

$$P_e \simeq \frac{2e^6 Z^2}{3m_e^2 c^3 b^4}. \tag{20}$$

The path of an electron flying past an ion is illustrated in Fig. 3. The time during which the Coulomb force is effective can be approximated by $2b/v$ where v

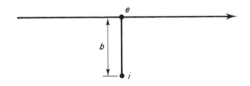

Fig. 3. Path of an electron flying past an ion; the impact parameter is b.

is the electron velocity and we shall regard the acceleration as constant during this time. The total energy radiated as the electron moves past is then

$$E_e \simeq \frac{4e^6 Z^2}{3m_e^2 c^3 b^3 v} \tag{21}$$

which is multiplied by the numbers n_e and n_i of electrons and ions per unit volume and by v to compute the rate of energy loss per unit impact area

$$P_a \simeq (4e^6 n_e n_i Z^2)/3m_e^2 c^3 b^2 \tag{22}$$

for all electron–ion collisions occurring in unit volume at an impact parameter b. The total power P_{br} radiated as *bremsstrahlung* per unit volume can now be obtained by multiplying by $2\pi b \, db$ and integrating from the distance of closest approach b_{min} to infinity. Thus

$$P_{br} \approx \frac{8\pi e^6 n_e n_i Z^2}{3m_e^3 c^3} \int_{b_{min}}^{\infty} \frac{db}{b} = \frac{8\pi e^6 n_e n_i Z^2}{3m_e^2 c^3 b_{min}}. \tag{23}$$

The minimum value of the impact parameter can be estimated by means of Heisenberg's principle

$$\Delta x \Delta p \simeq \frac{h}{2\pi}. \tag{24}$$

Setting the uncertainty of momentum equal to the electron momentum $m_e v$ and identifying Δx with b_{min} we obtain

$$b_{min} \simeq \frac{h}{2\pi m_e v}. \tag{25}$$

For a Maxwellian energy distribution we put

$$\tfrac{1}{2}m_e v^2 = \tfrac{3}{2}kT \tag{26}$$

and, substituting into (23) find

$$P_{br} \simeq \frac{16\pi^2}{3^{1/2}} \frac{(kT)^{1/2} e^6}{m_e^{3/2} c^3 h} n_e n_i Z^2. \tag{27}$$

A more precise derivation gives

$$P_{br} = g \frac{32\pi}{3^{3/2}} \frac{(2\pi kT)^{1/2} e^6}{m_e^{3/2} c^3 h} n_e \, \Sigma(n_i Z^2) \tag{28}$$

where g is the Gaunt factor which corrects the classical expression for quantum effects and is given by $2.3^{1/2}/\pi$ at high temperature. We have included an average over ions of different Z. Putting in the values of atomic constants

$$P_{br} = 5 \cdot 35.10^{-31} \, n_e \Sigma(n_i Z^2) \, T_e^{1/2} \text{ W cm}^{-3} \tag{29}$$

where T is now in keV. For deuterium ions only, $n_e n_i = n_D^2$, $Z = 1$, and for an equimolar mixture of D and T, $n_e = 2n_T$, $n_D + n_T = 2n_D$

$$P_{br} = 2 \cdot 14.10^{-30} \, n_D n_T T_e^{1/2} \text{ W cm}^{-3}. \tag{30}$$

It is instructive to inspect Fig. 4 where both the power output and the *bremsstrahlung* losses have been plotted for a plasma with $n_e = n_i = 10^{15}$. The pressure $p = (n_e + n_i)kT$ has also been plotted. This figure may be used for other densities by scaling the power with the square of the density ratio.

We must emphasize that impurities with high Z will greatly increase the radiation losses. For economic utilization, the densities must be chosen so that the output exceeds the radiation losses by a large factor.

We now pass on to the total energy loss of particles due to Coulomb interaction. Consider again Fig. 3, the momentum change of the electron in a direction perpendicular to the path is

$$I_\perp = \int E_\perp \, dt \tag{31}$$

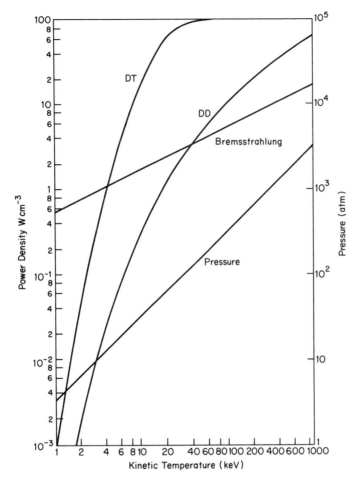

Fig. 4. Characteristics of thermonuclear reactions showing power output in W cm^{-3} vs temperature in keV and the power emitted as *Bremsstrahlung*.

where $E_\perp$ is the perpendicular component of the field. From Gauss's law $\int_{-\infty}^{\infty} \mathrm{d}x\, 2\pi b E_\perp = 4\pi Z e$. Hence

$$I_\perp = \int_{-\infty}^{\infty} E_\perp(t)\, \mathrm{d}t = \int_{-\infty}^{\infty} E_\perp(x) \frac{\mathrm{d}x}{v} = \frac{1}{v} \int_{-\infty}^{\infty} E_\perp(x)\, \mathrm{d}x = \frac{2Ze}{vb}. \qquad (32)$$

If $I_\perp$ is the momentum acquired by the electron the energy is $I_\perp^2/2m$. Integrating again over cylindrical shells of thickness $\mathrm{d}b$ we find the energy loss to be given by

$$-\frac{\mathrm{d}E}{\mathrm{d}x} = \frac{4\pi Z^2}{mv^2}\, n_e \ln \frac{b_{max}}{b_{min}}. \qquad (33)$$

Now we also need an estimate for b_{max}. The force on the electron may be considered as a pulse lasting a time $2b/v$ and to this time corresponds the frequency $v = v/2b$ of the dominant Fourier component. It will be seen in Chapter 4 that disturbances cannot propagate in a plasma if the frequency is inferior to the plasma frequency $v_p \approx 10^4 \sqrt{n_e}$. Thus we take the upper limit b_{max} as given by $b_{max} = v/2v_p$ or for relativistic electrons by

$$b_{max} = \frac{v}{2v_p(1 - \beta^2)^{1/2}}.$$

This gives

$$-\frac{dE}{dx} = \frac{4\pi Z^2 e^4 n_e}{mv^2} \ln \frac{mv^2}{2hv_p(1 - \beta^2)^{1/2}} \text{ erg cm}^{-1}. \tag{34}$$

It will have been noted that we have calculated the energy loss of electrons deflected by ions, but by kinematic inversion this is the same as the energy loss of ions on account of momentum imparted to electrons, but we must replace n_e by n_i. The loss of ions to other ions is small on account of the mass ratio.

Formula (34) is due to Bohr [10]. More precise calculations have been carried out by many authors, e.g. Bethe [11]. Brueckner and Jorna [12] give a formula which, for α-particles in the range of interest simplifies to give

$$-\frac{dE_d}{dx} = \frac{32\pi^{1/2}}{3} ne^4 \ln \Lambda \frac{E_\alpha^{1/2}}{T_e^{3/2}} \left(\frac{m_e}{m_\alpha}\right)^{1/2}. \tag{35}$$

Here log Λ is the so called Coulomb logarithm (Chapter 2, [14]) and may be taken to be $\ln b_{max}/b_{min}$ as given above. The range of a particle is defined as the integral

$$R = -\int_{E_0}^{0} \frac{dE}{f(E)} \qquad f(E) = -\frac{dE}{dx} \tag{36}$$

with a lower limit equal to the initial energy. According to Hughes and Schwartz [14], formula (36) gives

$$R_\alpha = \lambda_0(T^{3/2}/n),$$
$$\lambda_0 = 2 \cdot 10^{21} \text{ cm}^{-2}, \qquad T \text{ in keV}. \tag{37}$$

The fraction of the energy deposited is r/R_α for large R_α, and unity for $R_\alpha \to 0$.

For electrons a range cannot strictly speaking be defined. When absorption by a foil is measured, the number of electrons emerging from the foil steadily decreases with its thickness, but it is not possible to define a thickness s such that no electrons get through. Towards the end of the range a linear decrease is observed and an effective range is defined by extrapolation. Theoretical treatment is very difficult and empirical formulae have to be used. The matter is discussed in works on nuclear physics, e.g. [15], where data are given for three energy ranges. With the range R given in $g\,cm^{-2}$

$$R = 1.54\,E - 0.434 \qquad 1.2 < E < 2.3\,\text{MeV}$$
$$R = 1.1\text{--}63\,E - 0.36 \qquad 0.8 < E < 3\,\text{MeV} \qquad (38)$$
$$R = 1.099\,E^{1.38} \qquad 0.15 < E < 0.8\,\text{MeV}$$

These data refer to Al as an absorber and the effect of Z is discussed in [15].

For energies in the 10–100 range which is of interest in laser fusion, we quote tables compiled by Berger and Seltzer [13]. They compute the range by assuming that the electrons slow down continuously in an unbounded homogeneous medium from the initial to zero energy, assuming that the rate of energy loss along the entire track is equal to the mean rate. The data err on the high side but for our purpose, great accuracy is not needed. We find that an expression for the range in $g\,cm^{-2}$

$$R = 4.6(\log E/\log E_0)\,E^2, 0.1 > E > 0.01\,\text{MeV} = E_0 \qquad (39)$$

is accurate within 10%.

REFERENCES

1. Post, R. F. (1956). *Rev. Mod. Phys.*, **28**, 338.
2. Glasstone, S. and Lovberg, R. H. (1960). "Controlled Thermonuclear Reactions". Van Nostrand, Princeton, N.J.
3. Atkinson, R. d'E and Houtermans, F. (1928). *Z. Physik*, **54**, 656.
4. Gamov, G. (1938). *Phys. Rev.* **53**, 598.
5. Thompson, W. B. (1957). *Proc. Phys. Soc. (London)*, B **70**, 1.
6. Tuck, J. L. (1954). USAEC Report, LAMS-1640.
   Tuck, J. L. (1955). USAEC Report, LA-1190.
7. Wandel, C. F., Jensen, T. H., Hansen, O. K. (1959). *Nucl. Instr.* **4**, 249.
8. Eder, G. and Motz, H. (1958). *Nature*, **182**, 1140–1142.
9. Koch, W. and Motz, J. W. (1959). *Rev. Mod. Phys.* **31**, 29.
10. Bohr, N. (1913). *Phil. Mag.* **24**, 10.
    Bohr, N. (1915). *Phil. Mag.* **30**, 581.
11. Bethe, H. A. (1933). "Handbuch der Physik," Bd 24, 1, p. 518. Springer, Berlin.
12. Brueckner, K. A. and Jorna, S. (1974). *Rev. Mod. Phys.* **46**, 325–367.
13. Berger, M. G. and Seltzer, S. M. (1964). NASA report SP 3012 Nat. Aer. and Space Div. Washington, D.C.
14. Hughes, D. J. and Schwartz, R. B. (Eds) (1958). "Neutron Cross Sections", Brookhaven Nat. Lab., Report No 325 (US Government Printing Office) p. 58.
15. Segrè, E. (Ed.) (1953). "Experimental Nuclear Physics", John Wiley and Sons, New York, p. 297.

2

POWER GENERATION FROM FUSION REACTIONS

Using the information assembled in Chapter 1, it is now possible to review methods of generating electricity which use the output of fusion reactions. I do not intend to go into details of possible power stations, but rather to highlight the options which appear, and to comment on them in the light of existing knowledge.

A basic criterion for the possibility of a useful power generator was established by Lawson in 1957 [1]. It is concerned with the density n of nuclei per cm^3 of the fuel gas and the time τ during which the reaction must proceed at a given density in order to deliver a thermonuclear energy which is at least equal to the energy input. For a useful device, these minimum values of n and τ must, of course, be exceeded. This criterion states that the product of the number n of reacting nuclei per cm^3 and the time τ in seconds must be equal to a certain number $(n\tau)_{\mathrm{crit}}$ for "breakeven" and exceeds this value for a useful machine. This number is arrived at by the following consideration. Our generator must put energy into the electricity supply system but it also takes out energy in order to bring the reacting gas to the necessary temperature and density. In the case of laser fusion, the electricity from the supply is not very efficiently converted into light and a fraction of the light energy is absorbed by the reacting gas. Thus the energy W_p which is delivered to it is only a small fraction ε_p of the energy which must be taken out of the supply to keep the device working. Thus the energy generated by the device which is delivered to the supply system must exceed the energy W_p/ε_p taken from it. Let W_F be the fusion power; this is the heat which can be generated if all the energy E_F of the fusion reaction (17·4 MeV) (in the case of DT mixture) is turned into heat energy. It is given by

$$W_F = \langle \sigma v \rangle E_F n^2. \tag{1}$$

14

Let the efficiency of conversion of heat into electricity be ε_c, then we must have

$$\varepsilon_c \langle \sigma v \rangle E_F n^2 \tau + W_p \geq W_p/\varepsilon_p, \tag{2}$$

for breakeven the equality sign holds.

The reacting gas necessarily emits *bremsstrahlung*, and rate of which was calculated in Chapter 1 equation (19). If we take the view that this energy is lost, we must subtract it from the power output, but we may also suppose that it is turned into heat and recovered. A glance at Fig. 4 (Chapter 1) shows that the precise assumptions do not affect our estimate appreciably. In the case of the DT-reaction, for example, we see that, at a temperature of six MeV the fusion output is roughly twice the radiation loss and so, if we assume a temperature larger than six MeV there is only a factor of two involved.

The quantitative criterion also depends on whether the charged reaction products are or are not re-absorbed by the reacting gas. If they are, less heating power is required. If we assume that they are not, we take

$$W_p = n\theta_F \tag{3}$$

where θ_F is the particle temperature in keV. At temperatures in the keV region all the atoms are ionized and the reacting gas turns into a plama composed of ions and electrons. We shall discuss the properties of such a plasma in Chapter 4.

Assuming ε_p to be small compared to unity and $\varepsilon_c = \frac{1}{3}$, (2) becomes

$$\frac{1}{3}\langle \sigma v \rangle E_F n\tau \geq \theta_F/\varepsilon_p. \tag{4}$$

Taking a temperature of six keV we find from Fig. 1 (Chapter 1) that $\langle \sigma v \rangle = 3 . 10^{-17}$ so that, with $E_F = 17{\cdot}4 . 10^3$ keV

$$n\tau \geq (6/17{\cdot}4) \, \varepsilon_p^{-1} . 10^{14}. \tag{5}$$

Assuming that only half of E_F is available because the radiation energy is lost we get the criterion in the approximate form

$$(n\tau)_{\text{crit}} \geq \varepsilon_p^{-1} \, 10^{14} \, \text{cm}^{-3} \, \text{s}, \tag{6}$$

in which it is usually stated.

We shall now discuss this criterion in relation to various schemes of thermo-nuclear power generation. First we consider magnetic confinement. The reacting plasma is too hot to be contained by material walls, and it is therefore proposed to use a "magnetic bottle" i.e. a magnetic field configuration which curls up the particle orbits in such a way as to greatly reduce their escape rate from the reaction volume. Methods for doing this are discussed in detail in Chapter 1 [1, 2]. Speaking in a very general way, the confined state of plasma is far away from equilibrium, and there are many mechanisms leading to the break up, or dispersal, of the confined plasma. Research over many years has led to an understanding of these escape routes and many have been eliminated. There is, however, one phenomenon which is not yet understood; it was discovered by Bohm in 1949 [2]. According to him, there exists an anomalous diffusion process in a magnetic field B. The particles diffuse with a velocity given by

$$v_D = -\frac{ckT}{16en_eB}\nabla n_e = -\frac{8\cdot 64 . 10^3 \, T}{\gamma_B n_e B}\nabla n_e \qquad (7)$$

where γ_B is a factor which theory cannot account for in a straightforward way and which according to Bohm is 16 (see also [4, p. 47]). Much research in recent years has consisted of attempts to discover which circumstances are responsible for this anomalous diffusion, presumably caused by turbulent fluctuation processes.

Putting

$$\nabla n_e \approx n_e/r \qquad (8)$$

shows that Bohm diffusion limits the life time, $\tau \approx r/v_D$, of a plasma confined in cylindrical geometry with a radius r, to a value approximately given by

$$\tau = 10^{-4}(\gamma_B \gamma_i r^2 B)/T \qquad (9)$$

where γ_i is an improvement factor which may be realized by controlling the fluctuation level. Here B is measured in G, and T in K so that for $T = 10^8$ and a field of 100 kG and $r = 100$ cm, $\tau = 16$ ms if $\gamma_i = 1$.

Magnetic confinement systems are characterized by the ratio β of thermal energy $2nkT$ and magnetic energy $B^2/8\pi$,

$$\beta = 16\pi nkT/B^2. \qquad (10)$$

Tokomak is a low-β device with $\beta \approx 5\%$. The parameters of a possible reactor can be determined if we also consider the power loading of the reactor walls. A figure of 500 W cm^{-2} is commonly taken as a reasonable value. For plasma contained within a radius r the number N of particles per unit length is given [using (9)] by

$$N = \pi r^2 n = 10^4 \, \pi n\tau T/\gamma B \qquad (11)$$

and the power W reaching unit area of a wall of radius R is [using (10)]

$$W = \frac{3NkT}{2\pi R\tau} = \frac{3 . 10^4 \, B\beta T}{32 \, R\gamma}, \qquad \gamma = \gamma_B \gamma_c. \qquad (12)$$

Again for a temperature of $T = 10^8$ K we find from (12) that the wall radius must be at least

$$R = 5\cdot 9 \, \pi B\beta/\gamma. \qquad (13)$$

Now R must be larger than r and this condition yields

$$B \geqslant 3(\gamma/\varepsilon_p\beta^3)^{1/5} \text{ kG} \tag{14}$$

for the minimum value of B when r is determined from (9).

A reactor with $R = 100$ cm is already a very large device. Adopting this value and $B = 2.10^2$ kG which is perhaps practicable, we see from (13) that γ must be at least 1800 to satisfy our criteria. With this value and $\varepsilon_p = 0.01$, B must be larger than 180 kG which is consistent with our assumption of 200 kG. Such values of γ have not been achieved so far, even at temperatures smaller than thermonuclear ones, but it may well be possible to reach this goal. We have presented these estimates to show that the search for alternative methods for obtaining fusion power is still necessary.

The foregoing data imply confinement times of 1 ms and densities of 10^{18} per cm^3 which is again a large number, not easy to maintain for such a time or even longer.

It is for this reason that thoughts turned to microexplosions involving much higher densities during much smaller times. We speak of microexplosions because their effect must not be destructive. According to our criterion, with $\varepsilon_p = 0.01$, $n\tau = 10^{16}$, an explosion lasting 100 ps would correspond to a density of 10^{26} per cm^3 which would be just sufficient, provided that the fuel could be kept at this density during this time. Exploding matter spreads with a velocity exceeding the speed of sound, i.e. approximately $v = 1.5.10^8$ cm s^{-1} so that the size of a fuel pellet r_0 keeping together for a time $\tau = r_0/v$ is related to the density n_0 by

$$n_0 r_0 = 1.5.10^{22}\, \varepsilon_p^{-1}. \tag{15}$$

Hence, with the previous data, we arrive at a radius of 150 μm. A density of 10^{26} per cm^3 is approximately 2000 times solid density ($0.45.10^{22}$) of a DT mixture. Thus compression of the fuel pellet by a factor of the order of 10^4 is necessary in order to get a useful output.

It was suggested by Nuckolls et al. in 1972 [3] that the heating and compression might be accomplished by a powerful laser (see also [11, 12]). An electromagnetic wave will heat the electrons much faster than the ions. It is therefore very important to know whether in a time of 100 ps, the ion temperature can equilibrate with the electron temperature.

The energy exchange between hot and cold particle groups is treated in Spitzer's book [4]. Test particles of kinetic temperature T, mass m and charge Ze are supposed to enter the space filled by "field particles" with temperature T_f, mass m_f and charge $Z_f e$ and exchange energy by collision. It is found that the test particle temperature T changes according to

$$\frac{dT}{dt} = \frac{T_f - T}{t_{\text{equ}}} \tag{16}$$

where t_{equ} is given by

$$t_{equ} = \frac{3mm_f k^{3/2}}{8(2\pi)^{1/2} n_f Z^2 Z_f^2 e^4 \ln \Lambda} \left(\frac{T}{m} + \frac{T_f}{m_f}\right)^{3/2}$$ (17)

which may also be written

$$t_{equ} = 5.87 \frac{AA_f}{n_f Z^2 Z_f^2 \ln \Lambda} \left(\frac{T}{A} + \frac{T_f}{A_f}\right)^{3/2}$$ (17a)

in terms of the mass numbers A and A_f. Here $\ln \Lambda$ is the Coulomb logarithm which is discussed in Spitzer's book. For the purpose of our order-of-magnitude calculations it is a number of order 10. In the case of electrons considered as field particles and deuterons as test particles, $A = 2$ and $A_f = 2.82 . 10^{-4}$ and $t_{equ} = 0.7$ ps at a density of 10^{26} per cm^3 and $T_f = 10^8$ K ($T \ll T/A_f$). Thus there is enough time to heat the ions. We now turn to the problem of heating and compressing a pellet containing a deuterium–tritium mixture. We see it now as a spherical object with a radius of the order of 100 μm. The scheme as outlined by Nuckolls is to illuminate this target as uniformly as possible with intense laser light whereupon some light will be absorbed and material will evaporate from the pellet surface. At high temperatures it will ionize and form a rapidly expanding plasma mantle. Under intense illumination, the plasma density may rise to such an extent as to make it opaque to laser light. It is well known that light can only propagate if its frequency is higher than the plasma frequency, and we shall discuss its reflection and absorption in Chapters 7 and 8 in great depth. The heat from the absorption region will be conducted to the pellet where it will cause further evaporation. The pressure of the plasma so formed may become so high that a shock wave is launched into the target. The mass loss of the pellet by fast evaporation is called ablation. Under one-sided illumination the pellet would be accelerated like a rocket by the ablation process but under uniform illumination it will be compressed. The ablation process will be examined in Chapter 8, and Chapter 9 contains an introduction to the fluid dynamics of the compression process.

It is interesting to compare the conditions of a compressed pellet with those of hydrogen believed to exist in the sun at more than 10^3 times solid density and at pressures greater than 10^{11} atm. The compression work to be done on the target is resisted by the pressure of the dense matter often assumed to be represented by that of a Fermi-degenerate electron gas given by

$$p = \frac{2}{3} n_e \varepsilon_f \left[\frac{3}{5} + \frac{\pi^2}{4}\left(\frac{kT}{\varepsilon_f}\right)^2 - \frac{3\pi^4}{80}\left(\frac{kT}{\varepsilon_2}\right)^4 + \dots\right]$$ (18)

where n_e is the electron density and $\varepsilon_f = h^2/8m_e[(3/\pi)\,n_e]^{2/3}$ is the Fermi energy. At 10^4 times solid density, the minimum pressure occurs when $kT \ll \varepsilon_f$: it is 10^{12} atm. The question as to whether the pressure–density relation (18) (equation of state) is correct is examined in Chapter 11 which summarizes conclusions from theoretical work carried out in the astrophysical context.

It is important to reduce the laser power requirements to a minimum. The compression work needed depends on the temperature of the material before compression. It turns out, that in the process of laser-light absorption, "fast electrons", which can penetrate and heat the target before compression, are generated. If this "preheat" should become excessive, the laser power require ments would be too high for practical purposes.

We shall examine the experimental results regarding fast-electron gene- ration in Chapter 13. The problem of preheating is an important one and it determines target design and perhaps feasibility. The amount of fast electrons available for preheating the core, and their energy distribution, depends on the nature of the absorption process. It will be seen in Chapter 7 that there are several absorption processes, some of which involve fields which may accelerate electron groups, and high electron energies may result. The mean free path of fast electrons is approximately given by

$$\lambda_e\,(\text{cm}) = 6{\cdot}7.10^{-6}\,[E(\text{keV})]^2/\rho(\text{g cm}^{-3})$$

so that at $10\,\text{keV}$, $\lambda_e \approx 3.10^{-3}\,\text{cm}$ for $\rho = 0{\cdot}19\,\text{g cm}^{-3}$ (the density of solid DT) and such electrons penetrate deeply. On the other hand, this type of absorption may be needed in order to convert enough laser energy into compression work. The work done on the core by the fast electrons with suprathermal velocities of order V_{es} and pressure P_e which is turned into heat must clearly be small compared with the compression work P_iV_i done by the implosion wave with pressure P_i moving with the acoustic velocity $V_i = 10^6$ cm s^{-1} if the preheating is not to interfere seriously with the compression process. Assuming that $P = 1\,\text{Mbar}$, $P_e = 10^{-2}\,P_i$ and $V_{es} = 4.10^4\,\text{cm s}^{-1}$ ($10\,\text{keV}$) the ratio

$$\frac{P_e V_{es}}{P_i V_i} = f$$

becomes 4.10^{-4} and this may be deemed tolerable. On the other hand if $V_{es} = 10^5$, $P_e = 10^{-1}\,P_i$, the ratio is too high. It is hoped that suprathermal electron flux can be minimized by suitable target design. It can be seen from Fig. 7 of Chapter 13 that the electron energy increases with laser power density and it may therefore be necessary to design targets for the maximum tolerable f.

We will show in Chapter 9 that the maximum compression which can be effected by a single shock wave is limited, for a Fermi gas, to a 7-fold density increase. If the compression is to be accomplished by shock waves, a succession of shocks must be used. This requires the time history of the laser pulse to be controlled in such a way that successive shocks are generated and converge neatly at the target centre at the desired time. Programming of the time dependence of the laser pulse is not too difficult and has been achieved in several laboratories.

The compression process should be run as adiabatically as possible for minimum laser power requirements, that is, no heat should be lost on the way which does not serve the purpose of obtaining the ignition parameters of density and temperature. The word ignition is used to indicate that an important part of the heating can be accomplished by the absorption of the charged reaction products by the target. Formula (37) Chapter 1, shows that α-particles are in fact absorbed within a distance of 6 μm in a target of density 10^{26} per cm^3 when the temperature is 10 keV. It is also possible to arrange for the convergence of pulses in such a way, that the temperature of the centre of the core is sufficiently raised to start the reaction which then propagates with a supersonic burning front velocity. In an important review of the subject of laser fusion, Brueckner and Jorna (Chapter 1, reference 12) have treated various models of pellet heating analytically. They have also used their computation code to optimize parameters. Indeed, numerical codes are necessary for design work and we shall outline the basic physics underlying the Medusa code published by Roberts et al. [5]. Some examples of its application will be given in Chapter 12.

An analytic estimate of the fusion yields of cold DT fuel ignited by a spherically propagating burning wave has been given by Brueckner and Jorna. The result is presented in Fig. 1, showing the ratio of fusion energy to initial thermal energy in the pellet. The laser energy required to give this initial energy E_0 depends on the efficiency of laser energy deposition and fluid dynamic transfer to the compressed fuel. The authors suggest that it will be about 20 times larger than E_0. The curves are labelled by the compression ratio relative to solid DT ($0.19 \, g \, cm^{-3}$). To assess the useful energy multiplication we must know the efficiency of conversion of electricity into laser energy. If it is 1 %, a fusion yield of $10^4 \, E_0$ corresponds to an energy multiplication of five. This would require a 300-kJ laser. Larger multiplications can be obtained with even larger laser power.

One outstanding problem which has not so far been analysed in a satisfactory manner, is that of stability of the implosion. Chapter 10 is devoted to the study of stability, but only spatially uniform matter has been treated and the dynamics of the shock process have not been taken into account.

Performance can be much improved by good target design. Current thought is moving in the direction of hollow or multilayered targets, including denser materials, i.e. thin metal coating, to improve the coupling of laser power to the core; the targets currently used for experimentation are usually glass shells filled with DT gas. Results obtained with such targets are given in Chapter 13.

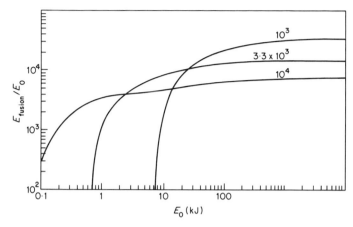

Fig. 1. Analytic estimate of fusion energy production in cold DT fuel ignited by a spherically propagating burning wave. The initial energy E_0 is the sum of ignition and degeneracy energy. The laser energy required to give this initial energy is larger by a factor which depends on the efficiency of transfer of laser power to the compressed fuel. It may be of the order of 20. The curves are labelled by the compression ratio relative to solid DT (0.19 g cm^{-2}.) (From [12] of Chapter 1, with permission.)

A tentative argument concerning the stability of an imploding thin shell runs as follows: it is shown in Chapter 10 that the growth rate γ of the Rayleigh–Taylor instability with wave number k for a large density ratio is given by

$$\gamma = \sqrt{kg}$$

where g is the acceleration and $k = 2\pi/\lambda$. In the plane case, λ is the wavelength of the rippling of a plane surface; in the spherical case such a rippling of the surface can be most conveniently analysed in terms of spherical harmonics, but the result is essentially the same.

Growing ripples then break up the symmetry of the implosion. The disturbance grows as $\exp \gamma t = \exp\sqrt{gk} = \exp\sqrt{gkt^2}$, but $\frac{1}{2}gt^2$ equals the radius R if t is the time the shock takes to reach the centre. Hence

$$\exp \gamma t = \exp\sqrt{2Rk}.$$

It is reasonable to assume that ripples with wavelengths smaller than the thickness ΔR of the shell are rather harmless, but break up the symmetry

B

when $\lambda > \Delta R$. We may thus assume that $k < 2\pi/\Delta R$ and so the dangerous ripples grow as $\exp(4\pi R/\Delta R)^{1/2}$. Opinions about allowable values for $R/\Delta R$ differ, a range is suggested by numerical work [6],

$$4 < R/\Delta R < 20$$

but precise predictions regarding the influence of heat-transport and viscosity cannot be made and the question of stability will be decided by experiment.

The compression of a DT target may also be accomplished by methods other than laser power. Linhart [7] suggested compression by imploding metal liners driven by conventional explosives. Kalisky et al. [8] report experiments in which a shock wave, driven by such explosive charges and repeatedly reflected by a linear was used to generate neutrons from a DT target.

It will be seen in Chapters 7 and 8 that the conversion of laser energy to compression and heating energy is a complicated process, and that it is not possible to predict the efficiency of the conversion. On the other hand, energetic charged particles may be used. Formula (34) in Chapter 1 shows that ions lose energy much faster than electrons. It has therefore been suggested that energetic ion beams could be used to initiate microexplosions instead of a laser. There appear to be two alternatives: the use of ion currents of many hundreds of kiloamps, with energies of several MeV or the use of low-ion currents of a few GeV. We shall first describe the basic features of a laser driven power station.

The following brief remarks are based on a thorough study compiled by L. A. Booth which he carried out at Los Alamos [9]. We consider the basic requirements of a laser-driven plant based on DT ignition. As soon as tritium is required, it is necessary to produce it artificially, and this can be done in the plant itself. Tritium is generated in a major fraction of reactions between lithium and neutrons. Lithium being a light element is also suitable for moderating high energy neutrons and converting them to thermal neutrons by elastic scattering processes. Thermal energy is also produced by neutron absorption in lithium. The reactor therefore consists of a large chamber capable of withstanding the effects of the microexplosion which contains a "lithium blanket" surrounding the target. The heat given up to the blanket is converted into electrical energy by means of a heat engine in a conventional thermodynamic cycle. A minor fraction of the fusion energy may also be directly converted by interaction with a magnetic field. The report envisages the target being surrounded by a porous wall through which lithium is transported from an outer reservoir to reach the evacuated explosion chamber. It is situated one metre from the centre of the spherical reactor vessel, and operates at 4000 K. It is heated by an energy release of 200 MJ from the DT burning, a quarter of which is deposited in the pellet while three quarters reach the blanket. Within milliseconds, the interaction of the pellet explosion products has subsided and kilograms of lithium have been vaporized. The hot gases escape through a supersonic diffuser and reach a condenser. The remaining heat is used in a heat exchanger through which liquid lithium from the reactor is circulated with an inlet temperature of 750 K. Such heat exchangers can be produced with existing technology. Reactor substances circulate in a closed cycle and the heat is conveyed to the engine plant. One duct for the admission of the laser beam is shown schematically. More inlets may be required and the beam may be deflected through several neutron-baffling bends by mirrors protected by liquid lithium layers. The structure must have a strong inner wall and the outermost shell may have a radius of two metres. The cycle of operations is repeated every second and the minimum power level is based on a thermal output of 200 MW, the concept calls for 10 modular units producing a total of 2000 MW. The reactor just described is illustrated in Fig. 2.

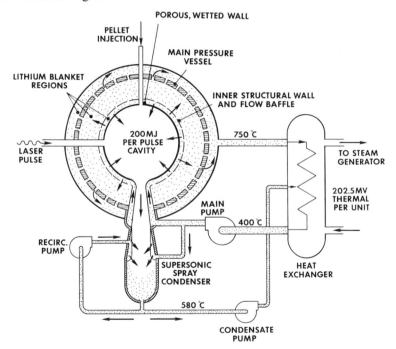

Fig. 2. Wetted wall concept of a laser fusion reactor. (From [9], with permission.)

The effect of the explosion is surprisingly small. The impulse associated with it is proportional to the square root of the product of the explosion energy and the mass of the explosion debris. Compared to chemical explosions producing the same energy, a fusion microexplosion has six orders of magnitude less debris mass, and thus an impulse, less by a factor of 10^3, to be absorbed by the reactor chamber. A 10^7-J fusion microexplosion produces no more impulse than a large firecracker. There is however a considerable blowdown from the wall caused by the arrival of the debris. The lithium cover of the porous wall protects it against the effects of the debris.

Other designs have been considered by other laboratories and KMS, a US firm very active in laser fusion research, has perfected a process in which neutrons are used to break up water molecules and produce hydrogen, or methane, which may then be piped through industrial or domestic gas supplies. The efficiency of this energy conversion is significantly higher than that attainable when electricity is produced and finally used for heating purposes.

In a study sponsored by ERDA an alternative system to laser fusion for delivering the power necessary for compression and ignition of a target was considered [10]. Heavy ions, e.g. Uranium ions, highly stripped, are accelerated and stored in a storage ring reaching a final energy of 10 MJ. The peak ion pulse of 600 TW and 10 ns duration is focused on the target, where an energy of 30 MJ is delivered with a repetition rate of 2 s to the target. The reactor chamber may be rather similar to the one described above. The accelerator and storage ring are of the type currently used by high-energy physicists. The study concludes that the method is economically feasible. One advantage is that the DD reaction may suffice without the use of tritium. Although there are still many open questions—e.g. the charge exchange cross-sections between heavy ions of an elevated temperature colliding in the storage ring are not well known—the study reaches optimistic conclusions and further research is being carried out in the USA. The low energy, high current version of particle fusion is not yet practicable because it is difficult to see how a comparable repetition rate can be reached with a focusable beam.

REFERENCES

1. Lawson, J. D. (1957). *Proc. Phys. Soc. (London)*, B **70**, 6.
2. Bohm, D. (1949). *In* "The Characteristics of Electrical Discharges in Magnetic Fields" (A. Guthrie and R. K. Wakerling, eds), Chapter 2, Section 5. McGraw Hill, New York.
3. Nuckolls, J., Wood, L., Thiessen, A. and Zimmerman, G. (1972). *Nature (London)*, **239**, 139.
4. Spitzer, L. (1962). "Physics of Fully Ionized Gases". Interscience, New York.
5. Christiansen, J. I., Ashby, D. E. T. F. and Roberts, K. V. (1974). *Comp. Phys. Comm.* **7**, 271–287.
6. Henderson, D. B. private communication.
7. Linhart, J. G. (1973). *Nucl. Fus.* **13**, 321.

8. Communication at the Oxford Conference on Laser Plasma Interact, Sept. 1977.
9. Booth, L. A. (1972). "Central Station Power Generation by Laser Driven Fusion", Los Alamos Report, LA 4858, MS vol. 1, 1 UC 20.
10. Erda summer study of heavy ions for inertial fusion (1976). Lawrence Berkeley Lab. and Lawrence Livermore Lab. LBL-5543.
11. Hora, H. (1969). *Annalen der Phys.* **22**, 402–404.
12. Basov, N. G., Krokhin, O. N. and Slitzkov, G. V. (1971). *In* "Laser Interaction with Plasma and Related Phenomena" (H. J. Schwarz and H. Hora, eds). Plenum Press, New York.

3

LASER SYSTEMS

Although this book is concerned with the physics of laser and particle fusion devices, its understanding will be helped by brief descriptions of the technological features of a high-power laser system. The reader is supposed to be familiar with low-power lasers, particularly the CO_2 laser and the neodymium (Nd) laser, the former pumped by electrical energy, the latter by light from flashlamps in order to achieve a population inversion of the lasing levels. The gain of a lasing system is theoretically treated in Chapter 5. The neodymium laser operates at a wavelength of $1\cdot06\,\mu m$, while the CO_2 laser furnishes power at a longer wavelength, $10\cdot6\,\mu m$. It will be explained in Chapter 8 that the coupling to the pellet core is better at the lower wavelength. The efficiency of CO_2 lasers is however much higher ($>2\%$) as compared with that of a neodymium laser and there are fewer problems connected with the amplification of high-power beams. At present, many experiments are carried out with the Nd laser, because of the more suitable wavelength. The problems of handling high-power beams do involve interesting physics and they will be explained.

We start by recalling that the radial intensity distribution of a beam being reflected froward and backward by the concave spherical mirrors of the resonator is Gaussian. A Gaussian field distribution is a solution of the wave equation

$$\nabla^2 v + k^2 v = 0, \qquad k = 2\pi/\lambda \qquad (1)$$

which is valid for field strength components; and the assumption of a wave

$$v = \psi(x, y, z)\, e^{-ikz} \qquad (2)$$

with an amplitude which varies slowly in the propagation direction, z

changes (2) to

$$\frac{\partial^2 \psi}{\partial x^2} + \frac{\partial^2 \psi}{\partial y^2} - 2ik\frac{\partial \psi}{\partial z} = 0. \tag{3}$$

It follows at once that

$$\psi = \exp\{-i(P + (k/2q))\, r^2\} \tag{4}$$

is a solution of (2) provided that $\partial P/\partial z = -i/q$ and that $\partial q/\partial z = 1$ and there-fore $q - z = $ constant. It is convenient to split $1/q$ into real and imaginary parts:

$$1/q = 1/R - i\lambda/\pi w^2. \tag{5}$$

Insertion into (4) reveals that $R(z)$ is the radius of curvature, i.e. the second derivative with respect to r, of the wave front that intersects the axis at z, so that all values of z are possible locations of mirrors with radius $R(z)$. Referring to the symmetry plane where the wavefront is plane ($R = \infty$) as the origin of the z-coordinate, $q_0 = i\pi w_0^2/\lambda$ and at a distance z

$$q = q_0 + z = i\pi w_0^2/\lambda + z. \tag{6}$$

Combining (5) and (6) and equating real and imaginary parts we find

$$w(z) = w_0[1 + (z/z_0)^2]^{1/2} \quad \text{and} \quad z_0 = \pi w_0^2/\lambda \tag{7}$$

and

$$R(z) = z[1 + (\pi w_0^2/\lambda z)^2]$$

Putting (5) into (4) shows that $\psi \propto \exp[-r^2/w(z)]$. Thus the beam profile is Gaussian and $w(z)$ may be defined as the beam radius.

Resonator design, and important beam properties are treated by Kogelnik and Li [1] in an excellent and very readable review. The book by Yariv [2] may be consulted on many topics concerning lasers. The book antedates the development of the iodine laser [3] which is also of importance for laser fusion.

Higher order modes (e.g. for x-polarization) with electric field components

$$E_{m,n}^{(x)} = E_0 H_m(\sqrt{2}x/w)\, H_n(\sqrt{2}y/w) \exp -(x^2 + y^2)/w^2$$

are also solutions of (2). Here H_m and H_n are Hermite polynomials of order m and n respectively. These higher modes are usually suppressed for laser fusion beams. For $m = n = 1$, the beam has the "Gaussian" profile $\exp(-r^2/w^2)$. For a discussion of higher order modes [1, 2] may be consulted. Formula (7) is important. It shows how the beam radius w varies with distance along the beam.

We shall discuss two physical processes, self-focusing and filamentation, because they determine the size and shape of amplifiers for solid-state (e.g. Nd) lasers. The history of the theory of self-focusing is interesting. Measured threshold values for the onset of stimulated Raman scattering in most liquids are far below the theoretically predicted values. This was explained by Askar'yian [4a] and by Talanov [4b] as being due to a small nonlinearity in the dependence of dielectric constant ε (or index of refraction n) on field strength. Indeed if

$$\varepsilon_{total} = \varepsilon + \varepsilon_2 E^2, \qquad \varepsilon_2 E^2 \ll \varepsilon \tag{8}$$

then

$$n(r) = n_0 + n_2 E^2 = \sqrt{\varepsilon_{total}} = n_0 + \frac{1}{2n_0}\varepsilon_2 E^2(r). \tag{9}$$

With E decreasing away from the axis of the beam, an off-axis "ray" will propagate faster, the further away it is from the axis, and an initially plane wavefront will become concave in the direction of propagation: the beam acts as its own focusing lens. Thus, the intensity of the beam in the Raman experiment was much increased due to self-focusing.

In mathematical terms, the wave equation for the amplitude E of the field of a linearly polarized light wave in a medium with dielectric constant (8) is given by

$$\nabla^2 E - 2ik\frac{\partial E}{\partial z} + k^2(\varepsilon_2/2\varepsilon)|E|^2 E = 0 \tag{10}$$

and by

$$\frac{1}{r}\frac{\partial E}{\partial r} - 2ik\frac{\partial E}{\partial z} + k^2(\varepsilon_2/2\varepsilon)|E|^2 E = 0 \tag{11}$$

in circular symmetry if $1/r(\partial E/\partial r) \gg \partial^2 E/\partial r^2$.

In one dimension (x or y) equations (10) is a nonlinear Schrödinger equation which turns up in various contexts. We shall meet it again in the chapters on nonlinear parametric processes, ponderomotive force and solitons. Whitham's book on nonlinear waves also treats self-focusing. (Chapter 9, reference [5]). To simplify a difficult problem we start from the solution of equation (11) with $\partial/\partial r \equiv 0$

$$E(z) = E(0) \exp[-i(k\varepsilon_2/4\varepsilon)|E|^2 z] \tag{12}$$

and modify it only in the exponent to describe initial focusing. Thus we write

$$E(r, z) = E(z) \exp[-i(k\varepsilon_2/4\varepsilon)|E(r, z)|^2 z] \tag{13}$$

and calculate the derivative appearing in (11)

$$\partial E/\partial r = (-ik\varepsilon_2/4\varepsilon)\, zE\, \partial/\partial r\, |E|^2. \tag{14}$$

We can write this term of (11) in another form by noticing that for a Gaussian beam with profile, $\exp(-r^2/w^2)$ at the beam edge, where $r = w$

$$-\frac{1}{r}\frac{\partial |E|^2}{\partial r} = -\frac{4}{w^2}|E|^2. \tag{15}$$

Looking at the resulting equation

$$\frac{\partial E}{\partial z} = \frac{\varepsilon_2}{2w^2\varepsilon}|E|^2\, zE - (ik\varepsilon_2/4\varepsilon)\,|E|^2\, E \tag{16}$$

we notice that the first term on the right-hand side corresponds to an intensity increase proportional to z while the second term is imaginary and therefore acts on the phase. We now define z, the focusing distance, as the distance in which the change of E is comparable to E and obtain

$$z_f \approx \frac{w}{|E|}\left(\frac{2\varepsilon}{\varepsilon_2}\right)^{1/2}. \tag{17}$$

Since the total power carried by the beam is $P \approx (\pi w^2 \varepsilon c)^{1/2}\, E^2$ we can also write

$$z_f = w_0^2\varepsilon\left(\frac{\pi c}{\varepsilon_2 P}\right)^{1/2}. \tag{18}$$

Neglecting transverse variations amounts to neglecting diffraction effects, and for this reason (16) predicts infinite intensity for $z \to \infty$, and there is no threshold for focusing. We can, however estimate the spreading of the beam due to diffraction from (7) and find the distance over which the beam radius doubles to be

$$z_d = \frac{\sqrt{3}}{\lambda}\,\pi w_0^2. \tag{19}$$

Equating this distance z_d to z_f we define the critical power P_c for self-focusing as

$$P_c = \varepsilon^2 c\lambda^2/(3\pi\varepsilon_2). \tag{20}$$

This may also be expressed in terms of the nonlinear refractive index $n_2 = \frac{1}{2}n(\varepsilon_2/\varepsilon)$ and becomes

$$P_c = \varepsilon_0 n^3 c\lambda^2/(6\pi n_2) \tag{21}$$

in mks units. It is useful to remember that $n_2(\text{esu}) = 9.10^8\, n_2$ (mks). Thus for CS_2, the substance used for the original Raman experiment for which $\varepsilon \approx \varepsilon_0$, $n_2\,(\text{esu}) = 10^{-11}$, $\lambda = 10^{-6}$ m, $P_c = \frac{4}{3}.10^4$ W. Thus self-focusing occurs at moderate power levels.

In the context of the discussion of Nd-laser systems, self-focusing is very important in that it limits the power density in the Nd glass rods of high power amplifiers. Indeed, self-focusing, if it leads to high intensity spots, destroys the (very expensive) rods. In order to cut down the energy density of the beam, large diameter glass rods (up to 20 cm!) are used. The glass must be of highest quality as far as homogeneity and freedom from internal stresses is concerned. It was found that the glass was shattered at power densities much smaller than those predicted by the foregoing theory. Again, Talanov provided the explanation; owing to the nonlinearity of the refractive index, a transverse modulation of the beam intensity grows so as to cause filamentation and a consequent intensity increase of the filaments which in turn causes the damage. The beam instability just characterized was analysed by Bespalov and Talanov along the following lines [5] (see also [6–9]).

The analysis again starts from equation (10) which, when $\partial^2 E/\partial z^2 \ll (\partial E/\partial z)\,k$ becomes

$$\nabla_\perp^2 E = 2ik\,\partial E/\partial z - k^2(\varepsilon_2/2\varepsilon)|E|^2\,E. \tag{22}$$

This is now solved on the assumption that a small perturbation which is periodic in the transverse (x, y) direction is superimposed on the smooth solution. Thus one writes

$$E = (E_0 + e)\exp[-i(k\varepsilon_2/4\varepsilon)|E_0|^2\,z] \tag{23}$$

where $e = e_1 + ie_2$ is a small perturbation assumed in the form

$$e_{1,2} = \text{Re}\{e_{1,2}^0 \exp i(\boldsymbol{\kappa}_\perp.\mathbf{r} - hz)\}. \tag{24}$$

This is substituted in (22) and only linear terms in e are retained. It is then found that h is imaginary in a certain parameter range, thus leading to unlimited growth of the disturbance. This method of stability analysis is the same as that used in the treatment of the Rayleigh–Taylor and Bénard instability given in Chapter 10.

The value found for h is

$$h = \frac{\kappa_\perp}{2}\left[2k\left(\frac{\varepsilon_2}{\varepsilon}\right)E^2 - \kappa_\perp^2\right]^{1/2}$$

which has its maximum

$$h_{\text{max}} = \Gamma = \tfrac{1}{2}k(\varepsilon_2/\varepsilon)\,E^2 \tag{25}$$

when

$$\kappa_\perp = \kappa_f = (\varepsilon_2/\varepsilon)^{1/2} kE. \tag{26}$$

The e-folding length for growth is therefore

$$\Lambda_\parallel = (\Gamma)^{-1} = \varepsilon_0 n^3 c/(4\pi n_2 P) \tag{27}$$

where P is the power density and the initial filament size

$$\Lambda_\perp = \frac{\pi}{\kappa_f} = n^{3/2} \lambda \sqrt{c\varepsilon_0}/[2(n_2 P)^{1/2}]. \tag{28}$$

Showing that the growth rate Γ is proportional to the laser intensity.

The instability pattern is initiated by any disturbance and it is therefore important to start with a "clean" beam profile. In particular an aperture causes a Fresnel diffraction pattern across the rod which starts up the instability. The remedy is to avoid sharp beam boundaries. One speaks of "apodizing" the beam. A combination of a suitably refractive liquid with a profiled rod constitutes such an apodizer which suppresses diffraction. Beam inhomogeneities can be removed by a spatial filter. The beam is focused on a diaphragm which cuts out spatial harmonics.

We have seen that the growth rate is proportional to intensity I. The tendency for beam break-up of the amplifier system is therefore measured by the so-called B-integral taken along the beam axis

$$B = \frac{2\pi\gamma}{\lambda} \int I \, \mathrm{d}l \tag{29}$$

where $\gamma = 4\pi . 10^7 . n_2/n_0 c$. Practical experience indicates that it must be kept smaller than a certain value, 4·2 for Nd rods.

In an amplifier rod, the intensity along the rod grows as

$$I = I_0 \exp(\alpha l) \tag{30}$$

and the B-integral is therefore inversely proportional to the amplifier gain α. We mention this fact because it shows that there is a limit to the rod diameter if an excessive growth of the B-integral is to be avoided. The following consideration shows how the limit arises. A glass rod of an amplifier is surrounded by flashlamps which supply the light which pumps the amplifying medium in order to obtain population inversion. The gain α is proportional to the inversion (see Chapter 5) which in turn is proportional to the pumping energy density. The pumping energy density is given by the ratio E_i/V of light input energy E_i to the volume V of the rod. But the input energy is proportional to the surface area of the rods and therefore this ratio is inversely proportional to the rod diameter. Altogether then, $\alpha \propto 1/r$ where r is the radius of the rod, and the B-integral is proportional to r.

We have learned that the B-integral which can be tolerated is limited, hence the rod diameter must not exceed a maximum value. It is for this reason that high-power disc amplifiers are used. Figure 1 shows a cross-section through a disc amplifier. Discs of neodynium glass are arranged to let the light enter and leave under the Brewster angle θ (the angle of incidence for zero reflection).

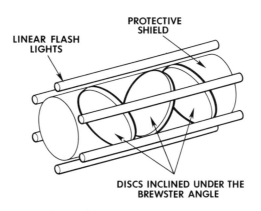

Fig. 1. Disc amplifier consisting of neodymium glass plates inclined under the Brewster angle, and flash lamps.

The discs have constant thickness and it is easy to see that now E_i/V is independent of r.

A laser system must also comprise components which prevent reflected power (particularly from the target) from travelling back, and provide feedback. Without them, superradiance would build up in the cavity formed between source and reflection plane and cause the oscillator to spill over before the desired pulse is shaped, and generally speaking the system as a whole would become unstable. One can use Faraday isolators, which are non-reciprocal transmission devices. They rotate the polarization plane in such a way that the double transit results in just that polarization which is suppressed by polarizing plates. Saturable dyes which do not transmit power below a certain intensity are used to define the oscillator power level.

The technology of CO_2 oscillators and amplifiers is undergoing rapid development. In their case, the pumping is done mostly by electrically driven gas discharges, or by the injection of electrons up to relativistic energies. High efficiency (up to 10%) has been achieved.

Having discussed the components which go towards the makings of a glass laser system, we now present a schematic characterization of an experimental installation (Fig. 2) which is taken from a report on the activities of the

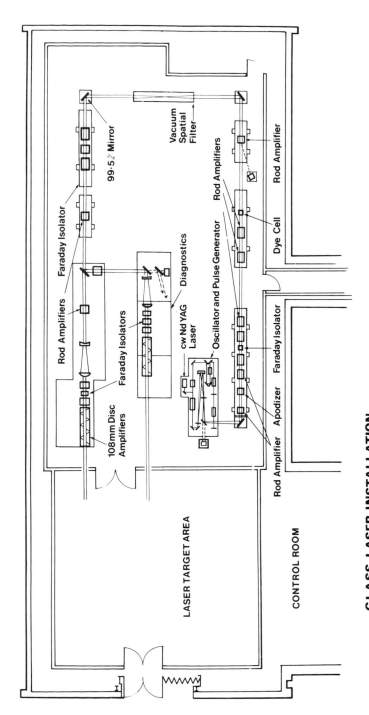

GLASS LASER INSTALLATION

Fig. 2. Glass laser installation (Rutherford Laboratory).

Rutherford Laboratory. It shows the arrangement of the oscillator, amplifiers, apodizer, Faraday isolators, dye cell, spatial filter, which produce two beams that enter the target area. An illustration of the arrangement in the target area is provided by Fig. 3. In the target chamber the target is seen to be illuminated by two symmetrically placed lenses. Around the target chamber, various instruments (the "diagnostic equipment") are deployed. The energy distribution of the ions emitted under illumination is measured, spectra of the

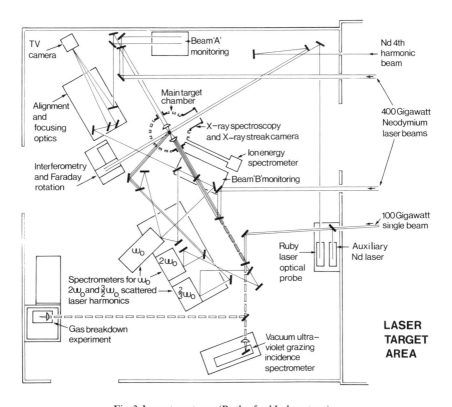

Fig. 3. Laser target area (Rutherford Laboratory).

light emitted by the plasma and added impurities are obtained, the emission of parametrically generated light at frequencies of $1\frac{1}{2}$ and 2 times that of the incident light as well as the reflected light is studied, and the X-ray emission is measured by various instruments. A pin-hole camera is used to obtain a picture of X-ray emission by the compressed target. We will show in Chapter 13 that it is also possible to obtain space-resolved and time-resolved pictures of X-ray emission during implosion.

The figures illustrate the kind of set-up required for experimentation. Much larger facilities exist in Limeil, France, in the Soviet Union and in Livermore and Los Alamos in the United States. We shall report some results obtained in the Lawrence Livermore laboratory and we list the code-words and performance figures in Table I.

Table I

	Janus	Cyclops	Argus
Number of beams	2	1	2
Aperture (cm/beam)	8·5	20	20
Max focusable power (TW)	0·4	0·7	6·0
Focusing optics	2 f/1	2 f/2·5	2 f/1
	2 f/0·47		
	ellipsoidal mirrors		lenses

REFERENCES

1. Kogelnik, H. and Li, T. I. (1966). *Appl. Opt.* **5**, 1550–1567.
2. Yariv, A. (1967). "*Quantum Electronics*". John Wiley and Sons, New York.
3. Witte, K., Braderlow, G., Fill, E., Hohl, K. and Volk, R. (1977). "Laser Interaction and Related Phenomena". Plenum Press, New York.
4a. Askar'yian, G. A. (1962). *Sov. Phys. JETP*, **15**, 1088.
4b. Talanov, V. I. (1964). *Radio Phys.* (transl.), **7**, 254.
5. Raspalov, V. I. and Talanov, V. I. (1966). *JETP*, **3**, 307.
6. Bliss, E. S., Speck, D. R., Holzrichter, J. F., Erkkilla, J. H. and Glass, A. J. (1974). *Appl. Phys Lett.* **25**, 448–450.
7. Campillo, A. J. and Shapiro, S. L. (1974). *Laser Focus*, June, 62–65.
8. Campillo, A. J., Shapiro, S. L. and Suydam, B. R. (1973). *Appl. Phys. Lett.* **23**, 628–630.
9. Campillo, A. J., Carpenter, B., Newnham, B. E. and Shapiro, S. L. (1974). *Opt. Comm.* **10**, 313–315.

4

BASIC NOTIONS CONCERNING PLASMA

Laser light impinging on a fusion target leads to the evaporation of the target material. When the light flux is high, the high-temperature gas ionizes quickly and soon consists of an electrically almost neutral mixture of ions and electrons: a plasma. It is almost neutral because departures from neutrality give rise to electric fields which restore it. This is shown by means of Poisson's equation

$$-\varepsilon_0 dE/dx = \varepsilon_0 d^2\phi/dx^2 = e(n_e - n_i) \tag{1}$$

where n_e and n_i are the electron and ion number-densities. A plasma with, for example—density of 10^{21} per cm^3 gives rise to an electric force of $1.6 \cdot 10^2$ V cm^{-1} when its neutrality is violated by 1% over the length of a micrometer. The conclusion that no potential differences can be maintained is however not justified because temperature motions oppose the restoring force. This may be illustrated by the example of a plasma consisting of electrons with charge $q = -e$ and hydrogen ions with charge $q = +e$. In equilibrium, at temperature T, the electron and ion densities in the presence of an electric potential ϕ will, according to Boltzmann's principle, be given by

$$n_e = n_0 \exp(e\phi/kT), \qquad n_i = n_0 \exp(-e\phi/kT). \tag{2}$$

Let the potential distribution be one-dimensionally non-uniform and depend on x in such a way that for $x \to \infty$, $\phi \to 0$, so that the density at infinity is n_0. Elsewhere the space charge density will be given by

$$\rho = (n_i - n_e)\, e = -2n_0 e \sinh(e\phi/kT). \tag{3}$$

The potential distribution is governed by Poisson's equation

$$\varepsilon_0 \frac{d^2\phi}{dx^2} = -\rho = 2n_0 e \sinh(e\phi/kT). \tag{4}$$

36

Equation (4) may be integrated to yield

$$\varepsilon_0 \left(\frac{d\phi}{dx}\right)^2 = 4n_0 kT[\cosh(e\phi/kT) - 1] \tag{5}$$

and this equation may be written in the form

$$(n_e + n_i)kT - \varepsilon_0 E^2/2 = 2n_0 kT \tag{6}$$

which is very instructive because it represents the stress balance between the sum of the partial particle pressures in the field free region appearing on the right-hand side and the algebraic sum of the particle pressures and the electrostatic stress appearing on the left-hand side.

It turns out that equation (5) is easily integrated once more to yield

$$\tanh(e\phi/4kT) = \tanh(e\phi_W/4kT)\exp(-|x|/\lambda_{D^2}), \quad \lambda_D = (\varepsilon_0 kT/n_0 e^2)^{1/2}, \tag{7}$$

where we have introduced a boundary condition putting $\phi = \phi_W$ at a constant potential wall at $x = 0$. The quantity λ_D is known as the Debye length and it is seen that, for $e\phi/kT \ll 1$ the potential dies out exponentially, and that is effectively screened out within a distance λ_D by the charges. Thus an electrode to which a potential is applied is surrounded by a sheath of a thickness of order λ_D leaving the rest of the plasma almost neutral or quasineutral. Small potentials of the order of kT/q still may exist but typically the ion concentration n_i is equal to n_e to better than one part in 10^6.

Numerically

$$\lambda_D = 6 \cdot 9\,(T/n)^{1/2}\,\text{cm} \qquad\qquad T \text{ in K}, \quad n = \text{number per cm}^3$$

$$\lambda_D = 7 \cdot 43 \cdot 10^2\,(T/n)^{1/2}\,\text{cm} \qquad T \text{ in eV}, \quad n = \text{number per cm}^3. \tag{8}$$

COLLECTIVE MOTIONS

It is remarkable that a combination of the Boltzmann equation and Maxwell's equations allows us to deduce most of the plasma properties which we shall need. What is so remarkable is that the Boltzmann equation which governs the space and time evolution of distribution functions f_e, f_i and f of electrons, ions and neutrals expresses only the simple fact that the number of particles of any of these species contained in a volume element $d\tau$ of phase space, the six-dimensional space spanned by the space and velocity coordinates, will only change if collisions between species remove particles from the volume element under consideration, or kick others, initially outside into it. Thus the equation for the distribution function in six-dimensional phase space

$f(x, y, z, w_x, w_y, w_z)$ specifying the number of particles in a volume element $dx\, dy\, dz\, dw_x\, dw_y\, dw_z$ centred on $\mathbf{r}$, a vector with components x, y, z and $\mathbf{w}$, the velocity vector with components w_x, w_y, w_z, is given by

$$df/dt = (\delta f/\delta t)_{\text{coll.}} \tag{9}$$

where the term on the right-hand side is the change of the number in the volume element brought about by collisions. The total time change of f, df/dt is given by

$$\frac{df}{dt} = \frac{\partial f}{\partial t} + \frac{\partial f}{\partial \mathbf{r}}\frac{d\mathbf{r}}{dt} + \frac{\partial f}{\partial \mathbf{w}}\frac{d\mathbf{w}}{dt} \tag{10}$$

but $d\mathbf{w}/dt = \mathbf{a}$, the acceleration which will be due to the Lorentz force $\mathbf{F}$ exerted by the electric and magnetic fields and $d\mathbf{r}/dt$ is, of course, the velocity $\mathbf{w}$, so that we can also write

$$\frac{\partial f}{\partial t} + \mathbf{w}.\frac{\partial f}{\partial \mathbf{r}} + \frac{q}{m}[\mathbf{E} + \mathbf{v} \times \mathbf{B}]\frac{\partial f}{\partial \mathbf{w}} = \left(\frac{\delta f}{\delta t}\right)_{\text{coll.}} \tag{11}$$

The distribution functions may be normalized by integrating over velocity space, setting

$$\int f(\mathbf{r}, \mathbf{w})\, d\mathbf{w} = n. \tag{12}$$

If the species have mean velocities, or drift velocities $\mathbf{v}$ they are given by the averages

$$\mathbf{v}(\mathbf{r}) = \langle \mathbf{w} \rangle = \int \mathbf{w} f(\mathbf{r}, \mathbf{w})\, d\mathbf{w} \Big/ \int f(\mathbf{r}, \mathbf{w})\, d\mathbf{w}. \tag{13}$$

The Boltzmann equation describes microscopic behaviour. From it, we can obtain the equations of fluid dynamics of performing averaging operations. By these operations we obtain moments of distribution functions. The lowest order average is performed by integrating equation (11) over velocity space. The next by multiplying equation (11) by $\mathbf{w}$ and again integrating over velocity space. This is done, e.g. in Spitzer's book (Chapter 2, reference [4]), and the first average leads to the continuity equation

$$\partial n/\partial t + \nabla. n\mathbf{v} = 0 \tag{14}$$

the second leads to the momentum equation

$$nm\, \partial \mathbf{v}/\partial t + nm(\mathbf{v}.\nabla)\mathbf{v} + \nabla.\boldsymbol{\Psi} + \mathbf{P} = n\mathbf{F} \tag{15}$$

where $\boldsymbol{\Psi}$ is an average over velocity space of the dyadic $\mathbf{uu}$

$$\boldsymbol{\Psi} = \langle nm\mathbf{uu} \rangle = m\int(\mathbf{w} - \mathbf{v})(\mathbf{w} - \mathbf{v})\, d\mathbf{w} = m\int \mathbf{uu}\, d\mathbf{w}. \tag{16}$$

Here $\mathbf{u} = \mathbf{w} - \mathbf{v}$, the difference between the particle velocity and its mean value is the random part of the velocity.

The quantity $\mathbf{P}$ is the momentum lost per unit time by collisions with other species arising from the right-hand side of equation (10) and $\mathbf{F}$ is the Lorentz force.

The average over velocity space of the quantity $\mathbf{u}$ vanishes because random velocities of opposite sign are equally probable. It is also assumed throughout that distribution functions and their velocity derivatives tend to zero for large values of the velocity, so that terms arising from partial integration vanish at $|\mathbf{w}| = \infty$. The quantity $\mathbf{\Psi}$ appearing in the momentum equation is not determined by (14) and (15) and the next higher moments are required to obtain an equation for its evolution. One says that the system of moment equations does not close. The quantity $\mathbf{\Psi}$ is the pressure tensor. (Pressure is defined as the rate of change of random particle momentum arriving at the boundary of a fluid element. Momentum, $m\mathbf{u}$ is transported with velocity $\mathbf{u}$). In rapid processes, collisions may not have time to equalize pressure in different directions of space and it can thus become anisotropic. When collisions are sufficiently frequent, the pressure is isotropic. In a suitable coordinate, the tensor $\mathbf{\Psi}$ becomes diagonal: we can write

$$\mathbf{\Psi} = \begin{vmatrix} \psi_{11} & 0 & 0 \\ & \psi_{22} & 0 \\ 0 & 0 & \psi_{33} \end{vmatrix}. \tag{17}$$

When the pressure p is isotropic $\psi_{11} = \psi_{22} = \psi_{33} = p$.

If it is desired to carry the approximation on to second moments, one can average the dyadic $\mathbf{ww}$. A new quantity, the heat flow tensor

$$\mathbf{Q} = m \int \mathbf{uuu}\, d\mathbf{w} \tag{18}$$

appears. (Heat flow is transport of random kinetic energy with the random velocity, a third-order quantity in $\mathbf{u}$.) One can break off the system here by assuming that div $\mathbf{Q} = 0$ and obtain an energy equation. We do not need it at this stage of our presentation.

The momentum equation (15) written in the form $nm\, d\mathbf{v}/dt = n\mathbf{F} - \nabla p$ equates the changes of momentum to the difference of the external force and the pressure gradient. In working with the momentum equation one needs the relation between pressure and density, the equation of state. For the purpose of studying plasma waves we shall assume an adiabatic relation

$$pn^{-\gamma} = \text{constant}. \tag{19}$$

When the plasma can be regarded as collisionless, the density changes due to the wave motion are rapid enough to make the pressure tensor anisotropic. It can then be shown, e.g. [2], that the adiabatic index is 3 and not $\frac{5}{3}$, the index valid for a neutral collisional monatomic gas. Under the influence of wave motion, positive and negative charges in a plasma separate. A polarisation takes place which has the spatial periodicity of the wave. This gives rise to spatially periodic electric fields, oscillating currents and magnetic fields. Such wave-like disturbances are collective effects, they involve a very large number of particles and give rise to macroscopic fields in addition to the microfields of the charged particles which vary on the time scale of their random motions. These collective effects are ordered in space and time and it is clear that external macroscopic fields will interact with collective motions. The simplest collective oscillations are longitudinal space charge waves in one dimension.

The plasma can also support a great variety of waves with electric and magnetic vectors parallel to perpendicular to the propagation direction $\mathbf{k}$ as well as hybrids of the two types. They are characterized by their polarization and by the dependence of the angular frequency ω on the wavelength λ or rather on the wave number $k = 2\pi/\lambda$. This dependence, the dispersion equation, will first be obtained for longitudinal waves.

We start from equations (14), (15) and see what happens when we perturb and equilibrium situation. In equilibrium, the pressure $p = nm\langle u^2 \rangle = nkT$. Writing $v = v_0 + v_1, n = n_0 + n_1, F = -eE_1$ where E_1, n_1 and v_1 are small quantities, we can neglect products of small quantities and obtain from (15)

$$n_0 \frac{\partial v_1}{\partial t} = -\frac{eE_1}{m_e} n_0 - \frac{1}{m_e} \frac{\partial p}{\partial x}. \tag{20}$$

The continuity equation gives

$$\frac{\partial n_1}{\partial t} = -n_0 \frac{\partial v_1}{\partial x}. \tag{21}$$

We assume that the ions are so heavy that they are at rest and only the electron density is disturbed. The charge excess is due to the electron density change n_1 and Poisson's equation becomes

$$\varepsilon_0 \frac{\partial E_1}{\partial x} = -en_1. \tag{22}$$

We want to eliminate v_1 by means of (21) to obtain an equation for n_1. We also want to replace the pressure gradient by a term involving the density gradient. This can be done by assuming the pressure change to be adiabatic according to (19) so that, with $p = nkT_e$

$$\frac{\partial p}{\partial x} = \gamma n^{\gamma - 1} \frac{\partial n}{\partial x} = \gamma \frac{p}{n} \frac{\partial n}{\partial x} = \gamma K T_e \frac{\partial n}{\partial x}. \tag{23}$$

Differentiating (21) with respect to t and using (22) we obtain

$$\frac{\partial^2 n_1}{\partial t^2} + e^2 n_1 n_0 / \varepsilon_0 m_e - \gamma \frac{kT_e}{m_e} \frac{\partial^2 n}{\partial x^2} = 0.$$

This has a solution with space time dependence $\exp[i(\omega t - kx)]$ which, when substituted leads to the equation

$$\omega^2 = \omega_p^2 + \gamma \frac{kT_e}{m_e} k^2, \qquad \omega_p^2 = \frac{e^2 n_0}{\varepsilon_0 m}, \tag{24}$$

expressing ω in terms of k, a dispersion equation. At very low temperatures, the electrons are seen to oscillate with the plasma frequency ω_p, but for finite temperature we get propagating waves. The oscillations come about because the charge disturbance produces an electric field proportional to the density, a restoring force similar to that produced by an elastic spring.

We have stated that for rapid one-dimensional compression, the adiabatic index becomes three and the dispersion equation becomes

$$\omega^2 = \omega_p^2 + 3 \frac{kT_e}{m} k^2. \tag{25}$$

At this point a word of caution is in order. Equation (25) is only valid if the phase velocity ω/k of the waves is large compared to the thermal velocity $(kT/m)^{1/2}$. This is due to a phenomenon discovered by Landau [4] and it comes about in the following way. If the wave moves with a speed of the order of the thermal speed some particles will travel in step with it and will interact continuously. It turns out that the wave may accelerate the particle and lose energy: the wave is then damped. Our treatment by means of the macroscopic equations cannot describe this behaviour of certain particle groups which get lost in the averaging procedure. It is necessary to use equation (11), the microscopic equation, together with Poisson's equation

$$\varepsilon_0 \frac{\partial E}{\partial x} = e \int du (f_i - f_e). \tag{26}$$

This combination first used by Vlasov [13] is usually called Vlasov theory.

Again we treat perturbations writing $f = f_0 + f_{1e}$ where f_{1e} is a small disturbance of the electron distribution assumed to vary as $\exp[i(\omega t - kx)]$. We neglect the motion of the heavy ions. Neglecting collisions, Boltzmann's equation (11) gives

$$\frac{\partial f_{1e}}{\partial t} + u \frac{\partial f_{1e}}{\partial x} = \frac{e}{m_e} E_1 \frac{\partial f_0}{\partial u} \tag{27}$$

where E_1 is assumed to vary as $\exp[i(\omega t - kx)]$. With this assumption

$$f_{1e} = -i \frac{e}{m_e} E_1 \frac{\partial f_0}{\partial u} / (\omega - ku) \tag{28}$$

and we insert this expression into equation (26) to obtain

$$i\varepsilon_0 k E_1 = -e \int du f_1 = -i \int \frac{e^2}{m_e} E_1 \frac{\partial f_0/\partial u}{(\omega - ku)} \tag{29}$$

so that the dispersion equation

$$1 + \frac{e^2}{\varepsilon_0 m_e k^2} \int_{-\infty}^{\infty} \frac{\partial f_0/\partial u \, du}{(V - u)} = 0 \tag{30}$$

results, where we have introduced the phase velocity $V = \omega/k$. The integral

$$I = \int_{-\infty}^{\infty} \frac{\partial f_0/\partial u}{(V - u)} du \tag{31}$$

requires careful discussion. When the phase velocity equals the particle velocity u there is a singularity. A complete discussion may be found in [2]. The meaning of the integral is explained by endowing the phase velocity with an imaginary component, writing

$$V = V_r \pm i\varepsilon V_i \tag{32}$$

and by approaching the real axis either from above or below, defining two integrals I_+ and I_- and going to the limit $\varepsilon \to 0$. The result of the analysis is that the solution of the dispersion equation depends on the sign of $\partial f_0/\partial v$. If the equilibrium distribution function f_0 has a positive slope, than for real k, ω has a negative imaginary part $\mathrm{Im}(\omega)$: the disturbance grows like $\exp[\mathrm{Im}(\omega) t]$ the plasma is unstable. If however $\partial f_0/\partial u < 0$ as it is in the tail of a Maxwell distribution, the waves are damped and decay as $\exp(-\gamma t)$ with γ given by

$$\gamma = \omega_p \sqrt{\pi} \left(\frac{m_e V^2}{2kT}\right)^{3/2} \exp\left(\frac{-m_e V^2}{2kT} - \frac{3}{2}\right) \tag{33}$$

$$V^2 = \frac{\omega_p^2}{k^2}$$

unless

$$V^2 \ll \frac{2kT}{m_e}.$$

This damping rate is known as the Landau damping rate. Landau [4] showed that, in the evaluation of the Cauchy integral (31) one has to write it as the sum of the principle part $P(V)$ and the contribution from the pole at $V = u$ which is $\pm i\pi \, \delta(V - u)$ where the positive sign has to be taken if $\partial f_0/\partial v < 0$, i.e. I_+ has to be used. He analysed the initial value problem, when the distribution function is given at $t = 0$. In this case the imaginary part of V must be positive, as demanded by the inversion of the Laplace transformation used for the treatment of the initial value problem [2, 3].

With the normalization $\int f_0(u) \, du = 1$ we can write

$$\varepsilon_L(\omega, k) = 1 + \frac{\omega_p^2}{k^2} \int \frac{\partial f_0/\partial u}{(V - u)} \, du = 1 - \frac{\omega_p^2}{\omega^2} \phi \left(\frac{\omega}{k}, f_0 \right). \tag{34a}$$

Introducing the variables

$$x = \frac{\omega}{\sqrt{2} k v_{Th}} \qquad v_{Th} = \left(\frac{kT}{m} \right)^{1/2}$$

and assuming a Maxwellian distribution of velocities $\varepsilon_L = 1 - \omega_p^2/\omega^2 \, \phi(x)$ with

$$\phi(x) = -2x^2 \left(1 - 2x \, e^{-x^2} \int_0^x e^{t^2} \, dt - i\sqrt{\pi} x \, e^{-x^2} \right) \tag{34b}$$

and ϕ may be expanded for $x \gg 1$ and $x \ll 1$ by noting that

$$2x \, e^{-x^2} \int_0^x e^{t^2} \, dt = 1 + \frac{1}{2x^2} + \frac{3}{4x^4} + \frac{15}{8x^6} \cdots, \qquad x \gg 1 \tag{34c}$$

$$= 2x^2 \left(1 - \frac{2x^2}{3} + \frac{4x^4}{15} \cdots \right), \qquad x \ll 1. \tag{34d}$$

For long waves, x is large. The real part of $\varepsilon_L = 0$ [i.e. equation (30)] gives

$$\omega^2 \approx \omega_p^2 [1 + 3(k\lambda_D)^2]$$

i.e. the Bohm and Gross equation (25) and setting

$$\omega = \omega (1 + i\gamma)$$

the imaginary part yields the Landau damping rate (33). For short wavelengths, one obtains

$$\varepsilon_L = 1 + (k^2 \lambda_D^2)^{-1} - i \left(\frac{\pi}{2} \right)^2 \frac{\omega_p^2}{\omega^2} \left(\frac{\omega}{k v_{Th}} \right)^2. \tag{34e}$$

Landau damping theory has many subtle points. The question arises as to how collisionless damping is possible. The result does not follow from the macroscopic equation and has therefore in the past been doubted by some authors. However, it has been confirmed in many experiments, and the intuitive explanation is that the wave attenuates by "phase mixing", that is by linear superposition of waves with different phase velocities. This implies, that the initial conditions in temporal Landau damping (or the boundary conditions in spatial damping) are important, and that not all initial data lead to exponential damping.

One can prove that the Vlasov equation conserves entropy. It follows that Landau damping, in the total absence of collisions, is a reversible process. There is a vast literature on the subject, and the reader interested in it will find information in [2, 3].

If the distribution function has a bump, as indicated in Fig. 1, which is a shape arising if an electron beam with a narrow velocity distribution is

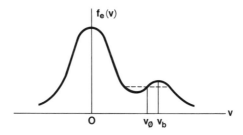

Fig. 1. Distribution function with a hump.

present, we have a region of positive slope which may make the plasma unstable. It may be obvious that the sign of γ is now reversed and that we obtain gain. Such a region may be looked at as a population inversion (there are more electrons with higher, and fewer with lower energy), see [5, p. 287]. The Landau gain is connected with population inversion, in the manner of the gain in an optically pumped active medium giving rise to laser action.

We shall also derive the Landau formula (33), which has just been obtained from small-signal linear theory, from the nonlinear theory of a later section.

At low frequencies the ion motion becomes important. We obtain the dispersion equation for ion acoustic waves by adding the ion susceptibility function to the electron term in (34a). The ion temperature T_i may be different from the electron temperature but $n_e = n_i$ and the ionic charge is Ze. When the proper expansions are inserted into the dispersion relation

$$\varepsilon_L = 1 - \frac{\omega_{pe}^2}{\omega^2}\,\phi(x_e) - \frac{\omega_{pi}^2}{\omega^2}\,\phi(x_i) = 0 \qquad (35)$$

one obtains for low frequencies

$$\frac{\omega}{k} = \left(\frac{kT_e}{m_i}\frac{1}{1 + k^2\lambda_D^2} + \frac{\gamma_i kT_i}{m_i}\right)^{1/2}.$$ (36)

It may be noted that the sound speed,

$$V_s = \left(\gamma\frac{p}{\rho}\right)^{1/2},$$

may be written

$$V_s = (\gamma kT/m_i)^{1/2}$$ (37)

for comparison with the ion sound speed for $\lambda_D^2 k^2 \ll 1$, $T_i \ll T_e$

$$V_{ia} \approx (kT_e/m_i)^{1/2}.$$ (38)

WAVE PROPAGATION IN A PLASMA

So far we have dealt with longitudinal waves only. We now want to study transverse waves of various types which the plasma can support. When the parameters are such that Landau damping does not take place, the fluid equations (14), (15), (19) can be used but they must be supplemented by Maxwell's equations

$$\nabla \times \mathbf{E} = -\mu_0\frac{\partial \mathbf{H}}{\partial t}$$ (39a)

$$\nabla \times \mathbf{H} = \varepsilon_0\frac{\partial \mathbf{E}}{\partial t} + \mathbf{j}$$ (39b)

$$\nabla.\mathbf{E} = \rho/\varepsilon_0$$ (39c)

$$\nabla.\mathbf{B} = 0.$$ (39d)

The current density $\mathbf{j}$ on the right-hand side of equation (39b) can be expressed in terms of the electric field by means of the relation

$$\mathbf{j} = \mathbf{T}\mathbf{E}$$ (40)

where, in general $\mathbf{T}$ is a tensor. This tensor character of $\mathbf{T}$ comes about because an oscillating magnetic field interacting with a charged particle with velocity $\mathbf{w}$ causes an oscillation in a direction perpendicular to $\mathbf{w}$.

We are interested in waves with space–time dependence, $\exp[i(\omega t - \mathbf{k}.\mathbf{n})]$

which, when substituted in equations (39) leads to

$$\mathbf{k} \times \mathbf{E} = \omega \mu_0 \mathbf{H} \tag{41a}$$

$$\mathbf{k} \times \mathbf{H} = -\omega \varepsilon_0 \mathbf{E} + i \mathbf{T} \mathbf{E} = -\omega \varepsilon_0 \left(\mathbf{I} + \frac{\mathbf{T}}{i\omega\varepsilon_0}\right) \mathbf{E} = -i\omega \varepsilon_0 \boldsymbol{\varepsilon} \cdot \mathbf{E} \tag{41b}$$

$$\mathbf{k} \cdot \mathbf{E} \quad = \rho/\varepsilon_0 \tag{41c}$$

$$\mathbf{k} \cdot \mathbf{B} \quad = 0. \tag{41d}$$

where $\mathbf{I}$ is the unit tensor and

$$(\mathbf{I} + \mathbf{T}/i\omega\varepsilon_0) = \boldsymbol{\varepsilon} \tag{42}$$

is a tensor dielectric constant. Substituting H from (41a) into (41b) we obtain:

$$\mathbf{k} \times \mathbf{k} \times \mathbf{E} = -\omega^2 \varepsilon_0 \mu_0 \boldsymbol{\varepsilon} \cdot E = -\frac{\omega^2}{c^2} \boldsymbol{\varepsilon} \cdot \mathbf{E}. \tag{43}$$

By means of the vector identity

$$\mathbf{k} \times \mathbf{k} \times \mathbf{E} = -k^2 \mathbf{E} + \mathbf{k}\mathbf{k} \cdot \mathbf{E} \tag{44}$$

this may be written in the form

$$-k^2 \mathbf{E} + \mathbf{k}\mathbf{k} \cdot \mathbf{E} + \frac{\omega^2}{c^2} \boldsymbol{\varepsilon} \cdot \mathbf{E} = 0. \tag{45}$$

We may take $\mathbf{k}$ to be in the z- (i.e. the 3-) direction and write

$$-k^2 E_1 + \frac{\omega^2}{c^2} (\varepsilon_{11} E_1 + \varepsilon_{12} E_2 + \varepsilon_{13} E_3) = 0 \tag{46a}$$

$$-k^2 E_2 + \frac{\omega^2}{c^2} (\varepsilon_{21} E_1 + \varepsilon_{22} E_2 + \varepsilon_{23} E_3) = 0 \tag{46b}$$

$$\frac{\omega^2}{c^2} (\varepsilon_{31} E_1 + \varepsilon_{32} E_1 + \varepsilon_{33} E_3) = 0. \tag{46c}$$

We have here three homogeneous linear equations of the coefficients which have a solution only when the determinant vanishes, i.e. when

$$\det \begin{vmatrix} -k^2 + \dfrac{\omega^2}{c^2} \varepsilon_{11}, & \dfrac{\omega^2}{c^2} \varepsilon_{12}, & \dfrac{\omega^2}{c^2} \varepsilon_{13} \\[2ex] \dfrac{\omega^2}{c^2} \varepsilon_{21}, & -k^2 + \dfrac{\omega^2}{c^2} \varepsilon_{22}, & \dfrac{\omega^2}{c^2} \varepsilon_{23} \\[2ex] \dfrac{\omega^2}{c^2} \varepsilon_{31}, & \dfrac{\omega^2}{c^2} \varepsilon_{32}, & \dfrac{\omega^2}{c^2} \varepsilon_{33} \end{vmatrix} = 0. \tag{47}$$

This then is the most general dispersion equation. An important special case arises if

$$\varepsilon_{31} = \varepsilon_{32} = \varepsilon_{13} = \varepsilon_{23} = 0,$$

and

$$\varepsilon_{12} = -\varepsilon_{21}, \qquad \varepsilon_{11} = \varepsilon_{22}.$$

This dispersion equation then has the form

$$\begin{vmatrix} W_{11} & W_{12} & 0 \\ W_{21} & W_{22} & 0 \\ 0 & 0 & W_{33} \end{vmatrix} = 0. \tag{48}$$

In this case the dispersion equation is satisfied if either

$$\begin{vmatrix} W_{11} & W_{12} \\ W_{21} & W_{22} \end{vmatrix} = 0 \tag{49}$$

(and this condition determines ω as a function of k for transverse waves) or

$$W_{33} = 0 \tag{50}$$

which establishes the relation for longitudinal waves. At this stage the equation is purely formal, and yet an important consequence may be deduced.

The equations for transverse waves are

$$W_{11}E_1 + W_{12}E_2 = 0 \tag{51a}$$

$$-W_{12}E_1 + W_{22}E_2 = 0. \tag{51b}$$

Thus one has

$$\frac{E_2}{E} = -\frac{W_{11}}{W_{12}} \quad \text{or} \quad \frac{E_2}{E_1} = \frac{W_{12}}{W_{22}}$$

and from (49) since $W_{11} = W_{22}$

$$W_{11}^2 + W_{12}^2 = 0 \tag{52}$$

and therefore

$$W_{11} = \pm i W_{12}$$

and

$$E_2/E_1 = \pm i. \tag{53}$$

Thus the component of $\mathbf{E}$ in the transverse direction have a phase difference of 90°, and the two signs refer to two waves with circular polarization with opposite sense of rotation.

It will now be shown how the conductivity tensor $\mathbf{T}$ and the tensor ε may be obtained from equations (14) and (15) and we shall briefly discuss the effect of a d.c. magnetic field $\mathbf{B}_0$. In our application, constant magnetic fields arise only under unfavourable circumstances, and we shall not go into great detail. We write the magnetic field as the sum of a constant magnetic field $\mathbf{B}_0$ and a small oscillating field $\mathbf{B}_1$

$$\mathbf{B} = \mathbf{B}_0 + \mathbf{B}_1. \tag{54}$$

Similarly we write the velocity as the sum of a constant drift velocity v_d and a small oscillating part v_1:

$$\mathbf{v} = \mathbf{v}_d + \mathbf{v}_1 \tag{55}$$

and the density as

$$n = n_0 + n_1. \tag{56}$$

When products of second order in the oscillating quantities are neglected the continuity equation (6) becomes

$$n_1(\omega - \mathbf{k} \cdot \mathbf{v}_d) = n_0 \mathbf{k} \cdot \mathbf{v}_1. \tag{57}$$

It is easy to show that, using the adiabatic law (17) the pressure term ∇p in the momentum equation (15) becomes

$$\nabla p = -in_0 \gamma kT \mathbf{k}(\mathbf{k} \cdot \mathbf{v}_1)/(\omega - \mathbf{k} \cdot \mathbf{v}_d). \tag{58}$$

We shall write

$$\omega_d = \omega - \mathbf{k} \cdot \mathbf{v}_d \tag{59}$$

which may be called a Doppler-shifted frequency which is the frequency measured by a stationary observer when a source moves with relative velocity v_d with respect to him.

The momentum equation becomes

$$iqn_0\mathbf{v}_1 = \frac{n_0 q^2}{m\omega_d} [\mathbf{E} + \mathbf{v} \times \mathbf{B}] + iqkn_0(\mathbf{k} \cdot \mathbf{v}_1)\frac{\gamma kT}{\omega_d^2} + q\frac{\mathbf{P}}{\omega_d}. \tag{60}$$

These equations may be written down for ions and electrons separately. We may also write them separately for various beams of particles with different drift velocities and regard them as separate species. Thus we supress the subscripts referring to different species and write the current density $\mathbf{j}$ as a sum over the species

$$\mathbf{j} = \Sigma q n \mathbf{v} = \Sigma q(n_0 + n_1)(\mathbf{v}_d + \mathbf{v}_1) = \Sigma q(n_0 \mathbf{v}_1 + n_1 \mathbf{v}_d). \tag{61}$$

To obtain the last equality we assume that the total d.c. current density $\Sigma q n_0 \mathbf{v}_d = 0$ vanishes. Similarly

$$\mathbf{v} \times \mathbf{B} = v_d \times \mathbf{B}_1 + v_1 \times \mathbf{B}_0. \tag{62}$$

We have put $\mathbf{v}_d \times \mathbf{B}_0 = 0$ because we shall asume that

$$\mathbf{v}_d \parallel \mathbf{B}_0. \tag{63}$$

The term $\mathbf{v}_d \times \mathbf{B}_1$ is small because, from Maxwell's equation (41) we obtain $|\mathbf{B}_1| = (k/\omega)|\mathbf{E}_1| = E_1/v_\phi$ and we assume $v_d/v_\phi \ll 1$. Neglecting, at first, the momentum transfer terms P and introducing the cyclotron frequency

$$\boldsymbol{\omega}_c = \frac{q\mathbf{B}_0}{m} \tag{64}$$

as a vector pointing in the $\mathbf{B}_0$ direction and the plasma frequency

$$\omega_p = \left(\frac{n_0 q^2}{m\varepsilon_0}\right)^{1/2}, \tag{65}$$

the momentum equation (60) becomes

$$qn_0v_1 - qn_0v_1 \times \boldsymbol{\omega}_c/i\omega_d - qn_0\mathbf{k}(\mathbf{k}\cdot\mathbf{v}_1)\frac{\gamma kT}{\omega_d^2 m} = \frac{\omega_p^2\varepsilon_0}{i\omega_d}\mathbf{E}, \tag{66a}$$

or, in components, since $\mathbf{k}\parallel z\parallel\omega_c$

$$qn_0v_{1x} - qn_0v_{1y}\frac{\omega_c}{i\omega_d} = \frac{\omega_p^2\varepsilon_0}{i\omega_d}E_x \tag{66b}$$

$$qn_0v_{1y} + qnn_{1x}\frac{\omega_c}{i\omega_d} = \frac{\omega_p^2\varepsilon_0}{i\omega_d}E_y \tag{66c}$$

$$qn_0v_{1z}\left(1 - k_z^2\frac{\gamma kT}{\omega_d^2 m}\right) = \frac{\omega_p\varepsilon_0}{i\omega_d}E_z. \tag{66d}$$

This may also be written

$$\mathbf{S}\, qn_0\mathbf{v}_1 = \frac{\omega_p^2\varepsilon_0}{i\omega_d}\mathbf{E} \tag{67}$$

by means of the tensor

$$
\mathbf{S} = \begin{vmatrix} 1 & -\omega_c/i\omega_d & 0 \\ \omega_c/i\omega_d & 1 & 0 \\ 0 & 0 & 1 - k_z^2\,\dfrac{\gamma kT}{m\omega_d^2} \end{vmatrix}. \tag{68}
$$

The reciprocal of $\mathbf{S}$ denoted by $\mathbf{S}^{-1}$ is easily found to be

$$
\mathbf{S}^{-1} = \begin{vmatrix} \omega_d^2/D & \omega_d\omega_c/iD & 0 \\ \omega_c\omega_d/iD & \omega_d^2/D & 0 \\ 0 & 0 & \left(1 - \gamma k_z^2\,\dfrac{\gamma kT}{m\omega_d^2}\right)^{-1} \end{vmatrix} \tag{69}
$$

where $D = \omega_d^2 - \omega_c^2$.

Thus the current density of equation (61) is given by

$$
\mathbf{j} = \Sigma(\mathbf{S}^{-1}\mathbf{E} + \mathbf{v}_d\mathbf{k}\cdot[\mathbf{S}^{-1}\cdot\mathbf{E}])\frac{\omega_p^2\varepsilon_0}{i\omega_d} \tag{70}
$$

and the conductivity $\mathbf{T}$ tensor defined by (40) through

$$
\mathbf{TE} = \Sigma\left[\mathbf{S}^{-1}\mathbf{E} + \frac{\mathbf{v}_d}{\omega_d}\mathbf{k}\cdot(\mathbf{S}^{-1}\cdot\mathbf{E})\right]\frac{\omega_p^2\varepsilon_0}{i\omega_d} \tag{71}
$$

is finally obtained and with it $\varepsilon = \mathbf{I} + \mathbf{T}/i\varepsilon_0\omega$. Note that the $\mathbf{S}^{-1}\cdot\mathbf{E}$ is a vector, hence $\mathbf{k}\cdot\mathbf{S}^{-1}\cdot\mathbf{E}$ is a scalar. Written as a tensor, $\mathbf{T}$ becomes

$$
\mathbf{T} = \Sigma\omega_p^2\varepsilon_0 \begin{vmatrix} \dfrac{\omega_d}{iD} & \dfrac{\omega_c}{D} & 0 \\[2mm] -\dfrac{\omega_c}{D} & \dfrac{\omega_d}{iD} & 0 \\[2mm] 0 & 0 & \dfrac{1 + uk/\omega_d}{i\omega_d(1 - \gamma kT/m\omega_d^2)} \end{vmatrix} \tag{72}
$$

where we have put $\mathbf{k}\|\mathbf{v}_d\|\mathbf{B}_0\|z$, $|v_d| = u$, $D = \omega_c^2 - \omega_d^2$.

We now return to the dispersion equation (48). The longitudinal part furnishes $W_{33} = 0$, i.e. $\varepsilon_{33} = (1 + T_{33}/i\omega\varepsilon_0) = 0$

$$
1 - \Sigma\frac{\omega_p^2}{\omega_d^2(1 - \gamma kT/m\omega_d^2)} = 0. \tag{73}
$$

When temperature effects are neglected we have

$$\Sigma \frac{\omega_p^2}{(\omega - uk)^2} = 1 \tag{74}$$

and for one species recover from (73) the Bohm and Gross equation (24) when $u = 0$.

It is interesting to note that we can recover equation (29). We assume that the different species under the sum are electron beams with continuously varying velocity, $\mathbf{u}\|\mathbf{k}$ with distribution function $f(u)$ and we replace the sum by an integral. The resulting equation

$$\frac{e^2}{\varepsilon_0 m} \int_\infty^\infty du f(u)/(\omega - uk)^2 = 1 \tag{75}$$

can be partially integrated (f vanishes for large u)

$$\frac{e^2}{\varepsilon_0 m k^2} \int_\infty^\infty du f(u)/(V - u) + 1 = 0$$

to lead back to (30) which we have already discussed.

We now briefly discuss the case of two electron beams with speeds $u_1 = u$ and $u_2 = -u$.

The dispersion equation

$$\frac{\omega_p^2}{(\omega - uk)^2} + \frac{\omega_p^2}{(\omega + uk)^2} = 1 \tag{76}$$

may be illustrated by Fig. 2. It is easy to see that the ω–k curves are hyperbolae with asymptotes $\omega = ku$, $\omega = -ku$. Equation (76) is a 4th-order equation, and therefore for each ω there must be four values of k, and vice versa. Therefore, in the middle region of the diagram where there are no real solutions, we must have complex solutions. They may represent growing or decaying waves.

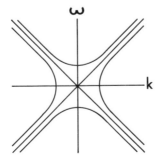

Fig. 2. Dispersion curves for counterstreaming cold electron beams with ions at rest.

The beam plasma system is unstable. Three types of instability may be distinguished: absolute, convective and evanescent. The third is not really an instability, it is pure spatial decay since spatial growth is not physically possible, for energetic reasons. It will be seen below that transverse waves decay exponentially in a plasma region with plasma frequency above the frequency of the wave, i.e. the wave does not penetrate deeply into what is called the overdense region. This is an example for evanescence. An absolute instability grows in time at a particular place: it is continuously fed by the beam at the expense of the energy of drift motion. A convective instability grows along the beam. It may start with some electrical noise which amplifies as it propagates.

Sturrock [6] was the first to show that it is possible by inspecting the dispersion curves, to determine the type of instability. We shall illustrate this by the simplest example.

The plasma can support waves with different dispersion characteristics, and, in order to study them, small terms of the equation which lead to coupling are neglected. However, when two dispersion curves cross, there is a resonance which brings small interaction terms into play. Crossing is avoided by reconnection of branches as illustrated by Fig. 3. Such a phenomenon is known in atomic physics when energy levels cross. Near the intersection the curves (by Taylor's Theorem) are second-order curves and where there is no real solution to the dispersion equation, there are again two complex ones. In Fig. 3a, there is a band of k-values for which there is no real ω. In Fig. 3b there is a band of ω values for which there is no real k, and in Fig. 3c there are ω-values for which there is no real k and k-values for which there is no real ω. It may be shown that Fig. 3a corresponds to an absolute instability, Fig. 3b to an evanescence and Fig. 3c to a convective instability. More complex intersections may lead to higher order curves and criteria have been discovered by Bers and Briggs see [7], which allow the determination of the type of instability.

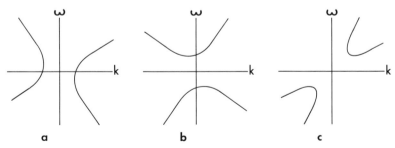

Fig. 3(a). Branches of dispersion curves corresponding to an absolute instability.
(b) Branches of dispersion curves corresponding to an evanescence.
(c) Branches of dispersion curves corresponding to a convective instability.

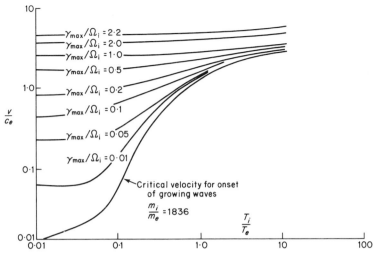

Fig. 4. Growth rate of the fastest growing wave in a current-carrying plasma. The normalized growth rates are constant along curves. Vertical axis: normalized drift velocity; horizontal axis: ratio of ion and electron temperatures. (From [8], with permission.)

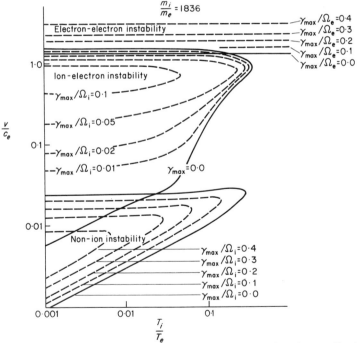

Fig. 5. Growth rates of the fastest growing wave in counter-streaming plasmas. Vertical axis: normalized relative velocity of two ion streams, or of ions streaming against electrons, respectively; horizontal axis T_i/T_e. (From [8], with permission.)

We have treated beam instabilities from the point of view of macroscopic theory. Stringer [8] has examined the problem from the point of view of microscopic theory i.e. Vlasov theory. The result can be summed up with reference to Figs 4 and 5. The information contained in these diagrams is very useful when one wishes to decide which type of instability is to be expected in a given situation. One very important instability is the double stream instability arising when ions and electrons drift with respect to each other.

Figure 4 shows contour lines of γ_{max}/Ω_i where γ_{max} is the maximum growth rate, i.e. the imaginary part of the frequency $\omega = \omega_i + i\gamma$, and Ω_i is the ion plasma frequency, in a plot with axes T_i/T_e and v/c_e where v is the beam velocity and c_e the electron thermal speed. The critical velocity which must be exceeded for instability to occur is also shown. Figure 5 is a similar plot for a counter-streaming hydrogen plasma showing the regions in parameter space where electron–electron, ion–electron and ion–ion instabilities occur when the respective beams of species move with relative velocity v.

The transition from cold- to hot-beam instability and the relation to Landau growth has been studied by Self et al. [9] and there is experimental work by Gentle et al. [10] on this topic.

Transverse Waves

We now return to the general dispersion equation (47) in order to extract information about transverse waves. The tensor **T** given by equation (72) may be specialized for the case when there is no d.c. magnetic field, i.e. $\omega_c = 0$ and no drift, i.e. $\omega_d = \omega$. We still take $\mathbf{k} \| \mathbf{z}$, $k_z = k$, so that for one species

$$\mathbf{T} = \omega_p^2 \varepsilon_0 \begin{vmatrix} \dfrac{i}{i\omega} & 0 & 0 \\ 0 & \dfrac{1}{i\omega} & 0 \\ 0 & 0 & T_{33} \end{vmatrix}$$

$\varepsilon = \mathbf{I} + \mathbf{T}/i\omega\varepsilon_0$.

We then obtain from (47) the dispersion equation for transverse waves

$$k^2 c^2 = \omega^2 - \omega_p^2 \tag{77}$$

which shows that for $\omega^2 > \omega_p^2$ there is propagation, but exponential decay

$$\exp\left[-(\omega_p^2 - \omega^2)^{1/2} z/c\right] \tag{78}$$

for

$$\omega < \omega_p.$$

This well known result explains a great wealth of phenomena, ranging from the colour of metals to ionospheric reflections. From our point of view it means that laser light will not penetrate much beyond the region of critical density in the plasma in front of the target where the laser frequency equals the plasma frequency.

Equation (47) with the tensor (72) contains much more information about waves in a magnetized plasma which, however, does not concern us closely in the context of laser fusion. It is concisely formulated in [2]. Full treatment can be found in [11, 12] and others.

We briefly touch the influence of collisions. The physics of collision processes is too complex to be taken into account in wave propagation problems. A simple plausible assumption is that the momentum transfer between electrons and neutrals P_{en} and the transfer between ions and neutrals P_{in} are proportional to the relative momentum on collision:

$$\mathbf{P}_{en} = -n_e v_{en} m_e \mathbf{v} \qquad \mathbf{P}_{in} = -n_i v_{in} m_i \mathbf{v}_i \qquad (79)$$

where v_{en} and v_{in} are the collision frequencies.

This modifies the tensor $\mathbf{S}^{-1}$ which, neglecting temperature effects becomes for electrons

$$i\omega \mathbf{S}_e^{-1} = \begin{vmatrix} \dfrac{1 + v_{en}/i\omega_{de}}{D_e} & \dfrac{\omega_{ce}/i\omega_{de}}{D_e} & 0 \\[2em] \dfrac{-\omega_{ce}/i\omega_{de}}{D_e} & \dfrac{1 + v_{en}/i\omega_{de}}{D_e} & 0 \\[2em] 0 & 0 & \left(1 + \dfrac{v_{en}}{i\omega_{de}}\right)^{-1} \end{vmatrix} \qquad (80)$$

$D_e = (1 + v_{en}/i\omega_{de})^2 - \omega_{ce}^2/\omega_{de}^2$ and with it, the conductivity tensor. Collisions between electrons and ions do not fit neatly into the theory developed above, because they couple the equations (60) written for electrons and ions. If ion motion is neglected v_{en} may be replaced by v_{ei} for a fully ionized gas.

A further comment on the plasma response will be useful later on. In Maxwell's equation (39b) we have put the current density equal to $\mathbf{TE}$. If an additional impressed current of density J is also present, equation (45) is modified to read

$$\left(\mathbf{kk} . -k^2 + \frac{\omega^2}{c^2} \varepsilon\right) \mathbf{E}_{\mathbf{k}, \omega} = -\mathbf{J}_{k, \omega} \mu_0 \omega i \qquad (81)$$

where $E_{k,\omega}$ and $J_{k,\omega}$ are Fourier components in space and time. This is now an inhomogeneous equation and we may now wish to calculate $E_{k,\omega}$ as a function of $J_{k,\omega}$. For this purpose we need the operator inverse $(\mathbf{kk} \cdot -k^2\mathbf{l} + (\omega^2/c^2)\boldsymbol{\varepsilon})^{-1}$ but this is a rather cumbersome expression and the following expedient may be adopted (e.g. Chapter 5, [3]).

The field is expressed as a product of a unit vector $\mathbf{e}_k$ in the polarization direction and its amplitude $E_{k\omega}$: $E_{k,\omega} = \mathbf{e}_k E_{k,\omega}$. Equation (45) is multiplied by $\mathbf{e}_k^*$ (we assume it complex for generality so that $\mathbf{e}_k \cdot \mathbf{e}_k^* = 1$), and we can now write formally

$$E_{k,\omega} = -i\omega\mu_0 J_{k,\omega} \cdot \mathbf{e}_k \left[(\mathbf{e}^* \cdot \mathbf{k})(\mathbf{e} \cdot \mathbf{k}) - k^2 + \frac{\omega^2}{c^2}(\mathbf{e}^*\boldsymbol{\varepsilon}\mathbf{e}) \right]. \qquad (82)$$

Russian authors put

$$\varepsilon_k^\sigma = \frac{1}{c^2}(\mathbf{e}_k^*\boldsymbol{\varepsilon}\mathbf{e}_k) + \frac{(\mathbf{k} \cdot \mathbf{e}_k^*)(\mathbf{k} \cdot \mathbf{e}_k)}{\omega^2} \qquad (83)$$

and the dispersion equation becomes (for the wave of type σ)

$$\frac{k^2}{\omega^2} = \varepsilon_k^\sigma(\omega, \mathbf{k}).$$

On inverting the Fourier transform the zeros of the dispersion equation become the poles of the integrand of

$$\frac{i}{(2\pi)^2} \int \int \frac{\omega\mu_0(\mathbf{e}_k \cdot J_k)\exp[-i(\omega t - \mathbf{kn})]\,d\mathbf{k}\,d\omega}{(k^2 - \omega^2\varepsilon_k^\sigma)} = \mathbf{E}(\mathbf{r}). \qquad (84)$$

We shall pursue this subject in connection with nonlinear waves and conclude this section with a comment on longitudinal dispersion equation $\varepsilon_{33} = 0$ (50). With the Vlasov treatment this reads

$$\left[1 + \frac{e^2}{mk^2\varepsilon_0} \int du \frac{\partial f_0}{\partial u} \middle/ (V - u) \right] E = 0$$

and

$$\varepsilon = 1 + \frac{e^2}{mk^2} \int du \frac{\partial f}{\partial u} \middle/ (V - u) \qquad (85)$$

is the dielectric constant in this case.

REFERENCES

1. Spitzer, L. (1962). "Physics of Fully Ionized Gases". Interscience, New York.
2. Motz, H. (1966). *Rep. Prog. Phys.* **29**, 623–674.
3. Allen, J. E. and Phelps, A. D. R. (1977). *Rep. Prog. Phys.* **40**, 1305–1368.
4. Landau, L. D. (1946). *J. Phys. USSR*, **10**, 25–34.
5. Bekefi, G. (1966). "Radiation Processes in Plasmas". Wiley, New York.
6. Sturrock, P. A. (1958). *Phys. Rev.* **112**, 1488–1503.
   Sturrock, P. A. (1960). *Phys. Rev.* **117**, 1426–1429.
7. Bers, A. and Briggs, R. J. (1963). Quart. Prog. Rep. No. 71, Electronics Lab., MIT, Cambridge, Mass.
8. Stringer, T. E. (1964). *Plasma Phys.* (*J. Nucl. Energy*, Part C), **6**, 267–79.
9. Self, S. A., Shouchi, M. M. and Crawford, F. W. (1971). *J. Appl. Phys.* **42**, 704–713.
10. Gentle, K. W. and Lohr, J. (1973). *Phys Fluids*, **16**, 1464.
11. Stix, T. H. (1962). "The Theory of Plasma Waves". McGraw Hill, New York and London.
12. Allis, W. P., Buchsbaum, S. J. and Bers, A. (1963). "Waves in Anisotropic Plasmas". MIT Press, Cambridge, Mass.
13. Vlasov, A. A. (1938). *J. Exp. Theor. Phys. USSR*, **8**, 291.

5

NONLINEAR PROCESSES

STIMULATED AND SPONTANEOUS TRANSITIONS

In this section we shall use the language of quantum physics, although the problems of plasma physics are essentially classical. This makes for great economy in the presentation of the basic laws. Planck's constant h is always eliminated in the final formulae. We shall treat nonlinear wave processes as spontaneous or stimulated emission or absorption of quanta. All wave are quantized, so we shall speak of plasmons, the quanta of longitudinal waves, of ionic plasmons, etc. We start with a digression about Eintsein's discussion of spontaneous and stimulated transitions in thermal equilibrium. [1].

We consider the interaction of an atom with black body radiation at temperature T whose energy density $\rho(v)$ per unit frequency v is given by Planck's law. We consider two energy levels of an atom and transitions between them. Let level 1 be higher than level 2 and consider the total transition rate w'_{12} from level 1 to level 2, in the presence of the radiation field

$$w'_{12} = B_{12}\rho(v) + A. \tag{1}$$

It consists of two parts, the spontaneous rate A and the rate induced by the radiation and proportional to its density: the stimulated transition. For energetic reasons, there can be no spontaneous upward transitions, so that the transition rate w'_{21} from level 2 to level 1 is taken as

$$w'_{21} = B_{21}\rho(v). \tag{2}$$

A has the dimension of reciprocal time and may be written $A = 1/t_{\text{spon}}$. B_{12} and B_{21} are constants to be determined. In equilibrium the rates must balance, hence if N_1 electrons are in the lower, and N_2 in the upper, state

$$N_2 w'_{21} = N_1 w'_{12}. \tag{3}$$

In addition the ratio N_1/N_2 by Boltzmann's principle is given by

$$\frac{N_1}{N_2} = \frac{g_1}{g_2} \exp(-hv/kT) \tag{4}$$

where g_1 and g_2 are the degeneracies (statistical weights) and hv is the difference between the energies of the two levels. From equations (1) and (2) we obtain

$$\frac{N_1}{N_2} = \frac{B_{21}\rho(v)}{B_{12}\rho(v) + A} \tag{5}$$

which can be solved for $\rho(v)$ using (4), with the result

$$\rho(v) = \frac{A(g_1/g_2)\exp(-hv/kT)}{B_{22} - B_{12}(g_1/g_2)\exp(-hv/kT)}. \tag{6}$$

Since the whole system, atoms and field, is in thermal equilibrium this must be identical with Planck's distribution

$$\rho(v) = \frac{8\pi hv^3}{c^3} \frac{1}{\exp(hv/kT) - 1} \tag{7}$$

and this is only possible if

$$B_{21} = B_{12}g_1/g_2 \tag{8}$$

and

$$\frac{A}{B_{12}} = \frac{8\pi hv^3}{c^3}. \tag{8'}$$

Thus according to (1) and (8′) the induced transition rate w_{12} due to interaction with the field is

$$(w'_{12})_{ind} = B_{12}\rho_v = \frac{Ac^3}{8\pi hv^3}\rho(v) = \frac{c^3}{8\pi hv^3 t_{spon}}\rho(v). \tag{9}$$

This result must apply to any radiation field with a "white" noise spectrum. In the case of monochromatic fields encountered in the application of these considerations to lasers we assume that the contribution at different frequencies are additive and write

$$(w'_{12})_{ind} = \int_{-\infty}^{\infty} \frac{Ac^3\rho(v')}{8\pi hv'^3} g(v' - v_0)\,dv' \tag{10}$$

where $g(v' - v_0)$ is a normalized line shape function. In a monochromatic field of frequency v,

$$\rho(v') = \rho_v \delta(v' - v) \tag{11}$$

and the transition rate becomes

$$w_{12} = \frac{c^3 A \rho_v}{8 \pi h v^3} g(v - v_0) = \frac{\lambda^2 I_v}{8 \pi h v t_{spon}} g(v - v_0) \tag{12}$$

which we have also written in terms of the incident energy flux $I_v = c\rho_v$, per unit area. We now calculate the absorption coefficient for monochromatic radiation in a medium with two levels populated with N_1 respectively N_2 electrons. The variation of energy flux due to stimulated transitions is

$$\frac{dI_v}{dt} = (N_1 w_{12} - N_2 W_{21}) hv = -\left(N_2 \frac{g_1}{g_2} - N_1\right) \frac{c^2 g(v - v_0)}{8 \pi v^2 t_{spon}} I_v \tag{13}$$

where we have more use of (12), (8).

The intensity will consequently vary as $I_v(z) = I_v(0) \exp(-\alpha(v) z)$

$$\alpha(v) = -\frac{1}{I_v} \frac{dI_v}{dz} = \left(N_2 \frac{g_1}{g_2} - N_1\right) c^2 g(v - v_0)/8 \pi v^2 t_{spon}. \tag{14}$$

It is immediately seen that the absorption coefficient α becomes negative, i.e. we obtain exponential amplification of the incident light instead of absorption of

$$N_1 > \frac{g_1}{g_2} N_2. \tag{15}$$

When this condition is satisfied, we speak of a population inversion. We have already indicated in our discussion of Landau damping that gain is associated with such an inversion.

Equation (13) is the basic equation for the gain in laser media.

WAVE–WAVE TRANSFORMATIONS

We now apply similar considerations to the emission and absorption of quanta of transverse electromagnetic waves and plasmons by a plasma. Plasmons, the quanta of plasma oscillations, are bosons which also obey Planck's distribution law. Now we find it convenient to work in terms of number N_k of quanta $h\omega$ at a point $\mathbf{r}$ in ordinary space with wave number $\mathbf{k}$ per unit volume in ordinary and k-space. Planck's law now becomes

$$N_k(\mathbf{r}) \, d\mathbf{k} = \frac{d\mathbf{k}}{\exp(\hbar\omega/kT) - 1} \qquad \hbar = \frac{h}{2\pi}, \quad \omega = 2\pi v \tag{16}$$

as shown in the section on radiation and absorption. Again we deal with any two of an infinite number of levels with occupation numbers n_1 and n_2 and write

$$\frac{\partial n_1}{\partial t} = -(a + l_{12}N_k)\, n_1 + l_{21}N_k n_2 = -\frac{\partial n_2}{\partial t}. \tag{17}$$

The equilibrium consideration given above must apply to any pair of levels and it shows that (for $g = g_1$)

$$a = l_{21} = l_{12} = \bar{w}. \tag{18}$$

Looking at equation (17) it is seen that the total probability for spontaneous and induced emission is

$$\bar{w}\,(N_k + 1)$$

while the probability for absorption is

$$\bar{w}N_k.$$

These are, of course, the matrix elements for transitions between oscillator levels familiar from quantum electrodynamics which treats the radiation field as an assembly of harmonic oscillators.

We apply this consideration to multi-quantum processes which arise when nonlinear problems are considered. So far, we have dealt with small wave amplitudes only, and we have neglected second order quantities like E_1^2, n_1^2, v_1^2, $n_1 v_1$. Because of such terms and even higher terms, the linear super-position law breaks down and waves are coupled. Transverse waves may be coupled to transverse and longitudinal waves. We speak of wave–wave transformation processes. In particular denoting transverse waves by t and longitudinal waves by l we shall have processes

$$t_1 \rightarrow t_2 + l.$$

The longitudinal waves may be ion acoustic waves or electron plasma waves, often called Langmuir waves. Such wave transformations are possible when momentum and energy are conserved. If a wave with label 0 is to be transformed into waves with labels 1 and 2 we must have,

$$\hbar\omega_0 = \hbar\omega_1 + \hbar\omega_2 \tag{19a}$$

and, in quantum language

$$\hbar\mathbf{k}_0 = \hbar\mathbf{k}_1 + \hbar\mathbf{k}_2 \tag{19b}$$

for momentum conservation.

Moreover, the values of ω and k for each wave must satisfy dispersion equations

$$\varepsilon_0(\omega_0, \mathbf{k}_0) = 0, \qquad \varepsilon_1(\omega_1, \mathbf{k}_1) = 0, \qquad \varepsilon_2(\omega_2, \mathbf{k}_2) = 0. \tag{19c}$$

It is useful and instructive to visualize the foregoing conditions by means of a geometric construction.† This is shown in Fig. 1 where three dispersion curves are drawn in an ω–k plot. The uppermost corresponds to that of electromagnetic waves, with its cut-off ω_p, that for Langmuir waves also has the same cut-off. The third curve corresponds to low frequency ion acoustic waves and almost fuses with the k-axis. The figure has rotational symmetry round the ω-axis so that what is shown is the intersection with the plane of the figure. The trick is to represent the frequency and propagation constant for

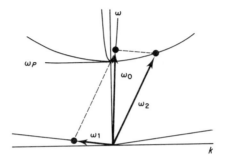

Fig. 1. Dispersion curves and k-vector diagram for wave–wave transformation.

each wave t, l, s (the letter s is used for ion-acoustic waves) as a vector, e.g. the vector $\mathbf{t}$ characterizing the wave t has ω_t as a component along the ω axis and k_t as its component along the k-axis. For the decay $t \rightarrow l + s$, $\mathbf{t} = \mathbf{l} + \mathbf{s}$ implies

$$\omega_t = \omega_l + \omega_s, \qquad \mathbf{k}_t = \mathbf{k}_l + \mathbf{k}_s. \qquad (20)$$

The dispersion equation is satisfied in the case of collinear k-vectors if the tips of the vectors lie on the respective dispersion curves. In general, 4-vectors are defined which have frequency as one component in addition to three components of $\mathbf{k}$ and their tips lie on the respective dispersion surfaces.

In order to compute the wave-transformation rates we shall apply the Einstein considerations to emission and absorption of waves by a plasma and calculate the rate of wave transformation, e.g. of a transverse wave t_0 into another transverse wave t_1 (of different ω and k) and a Langmuir wave l with $\mathbf{k}_l = \mathbf{k}_2$. In our formalisation, a wave (t_0) is absorbed and other waves (t_1 and l) are emitted. In the rate equation, a factor $1 + N_\mathbf{k}$ appears for an emission, and a factor $N_\mathbf{k}$ for an absorption process so that the probability

† This construction appears to have been used first by Peierls [2] in connection with solid state physics.

of decay of t_0 into t_1 and l will be given by

$$P_2 = \bar{w}N^t_{\mathbf{k}_0}(N^t_{\mathbf{k}_1} + 1)(N^l_{\mathbf{k}_2} + 1). \tag{21}$$

But we must also consider the inverse process in which a wave t_0 is emitted and waves t_1 and l are absorbed with probability

$$P_1 = \bar{w}(N^t_{\mathbf{k}_0} + 1)\,N^t_{\mathbf{k}_1}N^l_{\mathbf{k}_2}. \tag{22}$$

Thus the rate $dN^t_{\mathbf{k}_0}/dt$ will be given by $dN^t_{\mathbf{k}_0}/dt = \int d\mathbf{k}'_1\, d\mathbf{k}'_2 (P - P_2)$ or

$$\frac{dN^t_{\mathbf{k}_0}}{dt} = -\int \bar{w}\, d\mathbf{k}'_1\, d\mathbf{k}'_2 (N^t_{\mathbf{k}_0}N^l_{\mathbf{k}_2} + N^t_{\mathbf{k}_0}N^t_{\mathbf{k}_1} - N^t_{\mathbf{k}_1}N^l_{\mathbf{k}_2})/(2\pi)^6. \tag{23a}$$

Similarly

$$\frac{dN^t_{\mathbf{k}_1}}{dt} = \int \bar{w}\, d\mathbf{k}^t_0\, d\mathbf{k}'_2 (N^t_{\mathbf{k}_0}N^l_{\mathbf{k}_2} + N^t_{\mathbf{k}_0}N^t_{\mathbf{k}_1} - N^t_{\mathbf{k}_1}N^l_{\mathbf{k}_2})/(2\pi)^6 \tag{23b}$$

and

$$\frac{dN^l_{\mathbf{k}_2}}{dt} = \int \bar{w}\, d\mathbf{k}_0\, d\mathbf{k}_1 (N^t_{\mathbf{k}_0}N^l_{\mathbf{k}_2} + N^t_{\mathbf{k}_0}N^t_{\mathbf{k}_1} - N^t_{\mathbf{k}_1}N^l_{\mathbf{k}_2})/(2\pi)^6 \tag{23c}$$

neglecting terms of higher order in N.

Integrating each of these equations we find that

$$-\frac{d}{dt}\int N^t_{\mathbf{k}_0}\, d\mathbf{k}_0 = \frac{d}{dt}\int N^t_{\mathbf{k}_1}\, d\mathbf{k}_1 = \frac{d}{dt}\int N^l_{\mathbf{k}_2}\, d\mathbf{k}'_2. \tag{24}$$

These relations are known as the Manley–Rowe relations originally derived from purely classical consideration in a more complicated way [19]. Since each quantum carries an action h, each integral is proportional to the total action carried by the respective waves: the equations govern the change of action in the wave transformation process; and they express the fact that the number of quanta remains unchanged for slow processes (adiabatic invariance).

THE NUMBER OF QUANTA IN A RANDOM-PHASE WAVE PACKET

The number of waves with propagation vector $\mathbf{k}$ in an interval $d\mathbf{k}$ may be found by comparing the expression in terms of quanta

$$W = \int \hbar\omega_\sigma(\mathbf{k})N^\sigma_{\mathbf{k}}\, d\mathbf{k}/(2\pi)^3 \tag{25}$$

for the energy of a wave field σ with the equivalent expression

$$W = \frac{1}{2} \int_{-\infty}^{t} dt \left\{ E \frac{\partial D^*}{\partial t} + H \frac{\partial B^*}{\partial t} + \text{C.C.} \right\} \tag{26}$$

of a classical field. This is done in [3] with the result

$$N_{\mathbf{k}} = \frac{4\pi^3 \varepsilon_0 c^2}{\hbar \omega_\sigma^2} \left[\frac{\partial}{\partial \omega} \omega^2 \varepsilon_{\mathbf{k}}^\sigma \right]_{\omega = \omega_\sigma} |E_{\mathbf{k}, \sigma}|^2 \tag{27}$$

where $\varepsilon_{\mathbf{k}}^\sigma$ is defined by Equation (83) of Chapter 4, and for longitudinal waves by Equation (85) of Chapter 4.

THE TRANSITION PROBABILITIES

The transition probability $\bar{w}$ appearing in the rate equation may also be calculated by comparing classical and quantum expressions, this time for the radiation Q emitted per unit volume. Defining the probability $w_\sigma^{\sigma', \sigma''}$ for the transition from a wave type σ to waves of type σ' and σ'', the quantum expression is

$$Q_\sigma = \frac{\partial}{\partial t} \int \frac{\hbar \omega_\sigma N_{\mathbf{k}}^\sigma}{(2\pi)^3} \, d\mathbf{k} = \int \hbar \omega_\sigma w_\sigma^{\sigma' \sigma''} N_{\mathbf{k}}^{\sigma'} N_{\mathbf{k}}^{\sigma''} \frac{d\mathbf{k}' \, d\mathbf{k}'' \, d\mathbf{k}}{(2\pi)^9} \tag{28}$$

while the classical one is given by the work done by the current nonlinearly excited in the plasma by the waves σ' and σ'' on the wave field

$$Q_\sigma = - \int j(\mathbf{r}, t) E(r, t) \, dt. \tag{29}$$

This calculation is done for random phase waves in the book of Tsytovich [3] where the transition probabilities for many processes have been listed. We shall give an example of such a calculation.

ANALOGY BETWEEN NONLINEAR WAVE DECAY AND CHERENKOV RADIATION

Consider the process $t \rightarrow t + l$ illustrated in Fig. 2 by the k-vector diagram. From the figure one infers that

$$\cos \theta = \frac{\mathbf{k}_1^2 - \mathbf{k}_2^2 + \mathbf{k}_l^2}{2 \mathbf{k}_1 \mathbf{k}_l}. \tag{30}$$

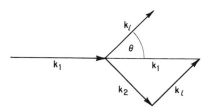

Fig. 2. k-Vector diagram for the decay of a transverse wave into another transverse wave and a longitudinal wave.

The dispersion equations for transverse waves Chapter 4, equation 77

$$\omega_{12}^2 = \omega_p^2 + \mathbf{k}_{1,2}^2 c^2, \tag{31}$$

together with the conservation equation $\omega_2 - \omega_1 = \omega_l$ allow us to transform (30) into

$$\cos \theta = \frac{\omega_l(\omega_1 + \omega_2)}{2\mathbf{k}_1\mathbf{k}_l c^2} + \frac{\mathbf{k}_l}{2\mathbf{k}_1}. \tag{32}$$

In the case when $\omega_1 \approx \omega_2$, $\mathbf{k}_1 \approx \mathbf{k}_2$, $\mathbf{k}_l \ll \mathbf{k}_1$ this becomes

$$\cos \theta = \frac{\omega_l}{\mathbf{k}_l} \cdot \frac{\omega_1}{\mathbf{k}_1} = v_\phi^l v_\phi \tag{33}$$

where we have introduced the phase velocities v_ϕ and v_ϕ^l of the electromagnetic and Langmuir waves. For the former we can prove from (31) that

$$v_\phi \cdot v_{gr} = c^2$$

and we obtain

$$\cos \theta = v_\phi^l / v_{gr}. \tag{34}$$

Thus the phase velocity of the Langmuir waves must be smaller than the group velocity of the t-waves, if such a decay is to take place. This is a result similar to the one obtained when material bodies emit waves. The velocity of a projectile must be smaller than that of the shock wave emitted and so is that of a charged particle emitting Cherenkov light. Here the group velocity of the light is analogous to the particle velocity.

CHERENKOV RADIATION

The Cherenkov effect may be explained with reference to Fig. 3 which shows the path of a charged particle travelling from A to a point B at distance l. Radiation emitted at A will, in propagating with phase velocity v_ϕ in the

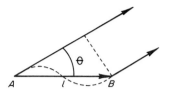

Fig. 3. Phase front formation by light emitted by electron in uniform or periodic transverse mor.

k-direction, be in phase with that emitted at B at a later instant in the same direction if

$$\frac{l}{v} - \frac{l\cos\theta}{v_\phi} = \frac{l}{v} - \frac{kl}{\omega}\cos\theta = 0. \tag{35}$$

In the direction θ determined by equation (35) a phase front will be formed, whereas in other directions the wave amplitudes cancel by interference. Thus the condition for emission is

$$\cos\theta = \omega/kv = v_\phi/v = c/n_r v \tag{36}$$

and the refractive index $n_r = c/v_\phi$ must be larger than unity since $v < c$. This is a special case of undulator radiation which arises when the particle path is periodic with a sideway motion of period l. We now can have

$$\frac{l}{v} - \frac{l}{c}\cos\theta = m\tau = \frac{2\pi m}{\omega}, \qquad m = 1, 2\ldots$$

and ω *in vacuo* is given by

$$\omega = \frac{2\pi mv}{l}\frac{1}{1 - v\cos\theta/c}. \tag{37}$$

The light can now be emitted *in vacuo* with $n_r = 1$ [4]. This process is important in connection with the free electron laser mentioned at the end of this book.

The condition for the emission of Cherenkov radiation will now be derived more rigorously from energy and momentum conservation.

Let the energy and momentum before emission be given by $\varepsilon_{p'}$ and $\mathbf{p}$ and by $\varepsilon_{\mathbf{p}}$ and $\mathbf{p}$ after emission. We must have

$$\mathbf{p}' = \mathbf{p} + \hbar\mathbf{k} \tag{38a}$$

$$\varepsilon_{p'} = \varepsilon_p + \hbar\omega. \tag{38b}$$

Putting the first into the second relation we obtain

$$\varepsilon_{(\mathbf{p}+\hbar k)} = \varepsilon_{\mathbf{p}} + \hbar\omega \tag{39}$$

When the wave momentum is small compared to the particle momentum

$$\varepsilon_{\mathbf{p}} + \hbar\omega \sim \varepsilon_{\mathbf{p}} + \hbar\mathbf{k} \cdot \frac{\partial\varepsilon_{\mathbf{p}}}{\partial\mathbf{p}} \tag{40}$$

and since even relativistically

$$\frac{\partial\varepsilon_{\mathbf{p}}}{\partial\mathbf{p}} = c^2\mathbf{p}/(p^2c^2 + m_0^2c^4)^{\frac{1}{2}} = \frac{c^2\mathbf{p}}{\varepsilon_{\mathbf{p}}} = \frac{\mathbf{p}c^2}{mc^2} = \frac{\mathbf{p}}{m} = \mathbf{v} \tag{41}$$

there results the previous condition

$$\hbar(\mathbf{k} \cdot \mathbf{v}) = \hbar\omega. \tag{42}$$

In our discussion of Landau damping according to linear theory, we could not take account of the change of the distribution function brought about by the particle oscillation because this leads to higher-order terms neglected in small-signal (linearized) theory. We shall now present a theory of wave emission by particles which yields the Landau result but also an equation for the change of the distribution function. This is called quasilinear theory and the equation in question is

$$\frac{\partial f_{\mathbf{p}}}{\partial t} = \frac{\partial}{\partial\mathbf{p}_i}\left(D_{ij}\frac{\partial f_{\mathbf{p}}}{\partial\mathbf{p}_j}\right) \tag{43}$$

which is an equation for diffusion in momentum space with D_{ij} as the diffusion constant. We shall also recover the Landau result in lowest order of approximation.

Consider a modified two-level system as illustrated by Fig. 4 with a difference $\hbar\omega$ between the energies of the levels and with electron energies distributed continuously so that the number of states of electrons with momentum $\mathbf{p}$ in the interval $d\mathbf{p}$ is given by $f_{\mathbf{p}}(d\mathbf{p}/(2\pi)^3)$. The probability of spontaneous emission of a wave σ by a particle with momentum $\mathbf{p}$ is $w_{\mathbf{p}}^{\sigma}$. The increase of the number $N_{\mathbf{k}}^{\sigma}$ of wave quanta due to emission is

$$\int (N_{\mathbf{k}}^{\sigma} + 1)w_{\mathbf{p}}^{\sigma}(\mathbf{k})f_{\mathbf{p}}\frac{d\mathbf{p}}{(2\pi)^3} \tag{44}$$

and the reduction due to absorption is

$$\int N_{\mathbf{k}}^{\sigma}w_{\mathbf{p}}^{\sigma}(k)f_{(\mathbf{p}-\hbar k)}\frac{d\mathbf{p}}{(2\pi)^3}. \tag{45}$$

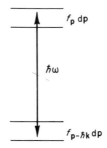

Fig. 4. Two-level system with continuous momentum distribution.

Since $\hbar k \ll p$

$$f_{\mathbf{p}-\hbar\mathbf{k}} \approx f_{\mathbf{p}} - \hbar\mathbf{k} \cdot \frac{\partial f}{\partial \mathbf{p}} \tag{46}$$

hence

$$\frac{dN_{\mathbf{k}}^{\sigma}}{dt} = \frac{\partial N_{\mathbf{k}}^{\sigma}}{\partial t} + \mathbf{v}_{\mathrm{gr}} \frac{\partial N_{\mathbf{k}}^{\sigma}}{\partial \mathbf{r}} = N_{\mathbf{k}}^{\sigma} \int w_{p}^{\sigma} \hbar\mathbf{k} \frac{\partial f_{\mathbf{p}}}{\partial \mathbf{p}} \frac{d\mathbf{p}}{(2\pi)^3} + \int w_{\mathbf{p}}^{\sigma}(k) f_{\mathbf{p}} \frac{d\mathbf{p}}{(2\pi)^3}. \tag{47}$$

We have included the space variation of the number of quanta and the appropriate velocity is the group velocity. Comparing this with the wave balance corresponding to (17),

$$\frac{\partial N_{\mathbf{k}}}{\partial t} = \tilde{w}N_{k}(n_1 - n_2) + \tilde{w}n_1$$

we see that the second terms correspond to spontaneous emission, and the first term is proportional to the population difference $n_1 - n_2$ which corresponds to $(df_{\mathbf{k}}/d\mathbf{p}) \, d\mathbf{p}$ in (47). Population inversion implies $n_1 > n_2$ or $df_{\mathbf{k}}/d\mathbf{k} > 0$. (Chapter 3, reference [5], p. 287.) For high intensities, we may neglect the spontaneous emission term and obtain the rate

$$\frac{dN_{\mathbf{k}}^{\sigma}}{dt} = \gamma_{\mathbf{k}}^{\sigma} N_{\mathbf{k}}^{\sigma} \tag{48}$$

with

$$\gamma_{\mathbf{k}}^{\sigma} = \int w_{\mathbf{p}}^{\sigma}(k) \hbar\mathbf{k} \frac{\partial f_{\mathbf{p}}}{\partial \mathbf{p}} \frac{d\mathbf{p}}{(2\pi)^3}. \tag{48a}$$

The growth rate is positive if the contributions from regions in momentum space where $df_{\mathbf{p}}/d\mathbf{p} > 0$ outbalance the contributions from the regions where $df_{\mathbf{p}}/d\mathbf{p} < 0$ which must occur since for large $\mathbf{p}$ the distribution function tends to zero.

We can now compute the change in the distribution function (for $N_k^\sigma \gg 1$)

$$\frac{df_\mathbf{p}}{dt} = \int [w_\mathbf{p}^\sigma(k)(f_{(\mathbf{p}+\hbar k)} - f_\mathbf{p}) - w_\mathbf{p}^\sigma(k)(f_\mathbf{p} - f_{(\mathbf{p}-\hbar k)})]N_\mathbf{k}^\sigma \frac{d\mathbf{k}}{(2\pi)^3}. \tag{49}$$

For example, the first term is the increase of the number of particles with momentum $\mathbf{p}$ due to transitions from levels with $p + \hbar k$, etc.

Substituting the Taylor expansion

$$f_{(\mathbf{p}\pm\hbar k)} = f_\mathbf{p} \pm \frac{\partial f_\mathbf{p}}{\partial \mathbf{p}_i}\hbar k_i + \frac{\hbar^2}{2}\frac{\partial^2 f_\mathbf{p}}{\partial \mathbf{p}_i\,\partial \mathbf{p}_j}k_i k_j \tag{50}$$

we obtain

$$\frac{df_\mathbf{p}}{dt} = \frac{\partial}{\partial \mathbf{p}_i}\left(D_{ij}\frac{\partial f_\mathbf{p}}{\partial \mathbf{p}_j}\right) \tag{51}$$

where

$$D_{ij} = \int \hbar^2 k_i k_j N_\mathbf{k}^\sigma w_\mathbf{p}^\sigma(k)\frac{d\mathbf{k}}{(2\pi)^3} \tag{52}$$

which is indeed the diffusion equation (43) announced before with a diffusion tensor D_{ik}. It is easy to show that particle energy plus wave energy is conserved as well as particle momentum plus wave momentum.

EXAMPLE FOR THE CALCULATION OF A TRANSITION

Probability and Landau Damping

The transition probability $w_\mathbf{p}^\sigma$ for the emission of wave σ by an electron of momentum p can be determined by comparing the quantal expression for the energy radiated per unit of time

$$W_\mathbf{p}^\sigma = \int \hbar\omega^\sigma w_\mathbf{p}^\sigma \frac{d\mathbf{k}}{(2\pi)^3} \tag{53}$$

with that calculated by working out the work done by the electric force of the wave radiated on the charge in motion which produces it. The field is found from Chapter 4, equation (82) to be

$$E_{\mathbf{k},\,\omega} = \frac{-i\mu_0\omega(j_{\mathbf{k},\,\omega}\cdot e_\mathbf{k}^\sigma)}{(k^2 - \omega^2\varepsilon_\mathbf{k}^\sigma)}. \tag{54}$$

The Fourier component $\mathbf{j}_{k,\,\omega}$ of a charge moving in rectilinear motion $\mathbf{r} = \mathbf{v}t$

is given by

$$\mathbf{j}_{k,\omega} = \int e\mathbf{v}\delta(\mathbf{r} - \mathbf{v}t)\exp\left[i(\omega t - \mathbf{k}.\mathbf{r})\right]\mathbf{dr}\,dt = 2\pi\,e\mathbf{v}\delta(\omega - (\mathbf{k}.\mathbf{v})) \quad (55)$$

The rate of work by the Fourier component of radiation integrated over all $\mathbf{k}$ and ω becomes

$$W = \int \mathbf{j}^*\mathbf{E}\,\mathbf{dr} = \int \frac{ie^2\mu_0\omega[\mathbf{v}.\mathbf{e}_\mathbf{k}^\sigma]^2\delta(\omega - \mathbf{k}.\mathbf{v})\delta(\omega' - \mathbf{k}'.\mathbf{r})}{(2\pi)^6(k^2 - \omega^2\varepsilon_\mathbf{k}^\sigma)} \times$$

$$\exp\left[i(\mathbf{k} - \mathbf{k}').\mathbf{r}\right]\exp\left[i(\omega - \omega')\right]\mathbf{dr}\,d\omega\,d\omega'\,\mathbf{dk}\,\mathbf{dk}'. \quad (56)$$

Only the imaginary part of the denominator of $E_{\mathbf{k},\omega}$ contributes. We put

$$\mathrm{Im}\,(\omega^2\varepsilon_\mathbf{k}^\sigma)^{-1} = \pi\left(\frac{\partial}{\partial\omega}\,\omega^2\varepsilon_\mathbf{k}^\sigma\right)^{-1}_{\omega=\omega^\sigma}\left[\delta(\omega - \omega^\sigma) + \delta(\omega + \omega^\sigma)\right] \quad (57)$$

where ω^σ is a solution of eq (83) of Chapter 4, which makes

$$W = -\int e^2\pi\frac{\mu_0\omega\delta(\omega - \mathbf{k}.\mathbf{v})(\mathbf{v}.\mathbf{e}_\mathbf{k}^\sigma)^2\delta(\omega - \omega^\sigma)}{\partial/\partial\omega\,(\omega^2\varepsilon_\mathbf{k}^\sigma)}\frac{\mathbf{dk}\,d\omega}{(2\pi)^3} = -W_\mathbf{p}^\sigma \quad (58)$$

real.

Comparing (58) with (53) we find

$$w_\mathbf{p}^\sigma = \frac{\mu_0 e^2\pi}{\hbar}\frac{(\mathbf{v}.\mathbf{e}_\mathbf{k}^\sigma)^2}{\partial/\partial\omega\,(\omega^2\varepsilon_\mathbf{k}^\sigma)_{\omega=\omega^\sigma}}\delta(\omega^\sigma - \mathbf{k}.\mathbf{v}). \quad (59)$$

The number of quanta $N_\mathbf{k}^\sigma$ also has h in the denominator so that it drops out in (48a) and (52). The result is classical. It is easy to calculate the result

$$W_p^t = e^2v\int\omega\,d\omega\left(1 - \frac{v_\phi^2}{c^2}\right)$$

first obtained by Frank and Tamm for the energy radiated as transverse waves (Cherenkov radiation) per unit time. For plasma oscillations with

$$\omega \approx \omega_{pe}, \qquad \varepsilon = 1 - \omega_{pe}^2/\omega^2, \qquad \mathbf{e}_k^l = (\mathbf{k}/k),\,\mu_0 \to 4\pi$$

we obtain

$$w_\mathbf{p}^l = \frac{4\pi^2 e^2\omega_{pe}}{\hbar k^2}\delta(\omega_{le} - \mathbf{k}.\mathbf{v}). \quad (60)$$

Substituting (60) into (48) and taking a Maxwellian distribution

$$f_p = \left(\frac{m}{2kT\pi}\right)^{1/2} \exp(-v^2/2v_{Te}^2), \qquad \omega_l^2 = \omega_p^2 + \tfrac{3}{2}k^2 v_{Te}^2, \qquad v_{Te} = \sqrt{2kT/m}$$

$$\tag{61}$$

for the electrons, we find, by an elementary integration

$$\gamma_k^l = -\sqrt{\frac{\pi}{8}}\, \omega_{pe} \frac{\omega_{pe}^3}{k^3 v_{Te}^3} \exp\left[-(\omega_{pe}^2/2k^2 v_{Te}^2) - \tfrac{3}{2}\right] \tag{62}$$

which is the Landau result in an equivalent notation. Note that damping occurs because $\partial f_p/\partial p$ in formula (48a) is now negative.

With the help of the quasilinear equation (51) for the change of the distribution function one can now study the saturation of Landau damping: the distribution flattens out, reducing df_p/dp. The subject of nonlinear Landau damping is however quite complicated when further wave–wave processes are considered. We shall come back to it in a section on particle trapping.

SCATTERING PROCESSES IN A PLASMA

We have just described stimulated emission and absorption of waves by particles. Now we want to relate these processes to scattering processes. First consider scattering in the absence of plasma effects. A wave may be scattered by Thomson, or Compton scattering: the wave field makes the particle carry out oscillating motions, and as a consequence, dipole radiation is emitted by the particle. In the elementary process, a quantum of radiation, is absorbed and another one emitted as illustrated in Fig. 5(a) where the full line is the particle path and the dashed lines are the wave-path.

To the extent that the particle loses energy, the wave frequency is changed to conserve energy. In a plasma, more complicated nonlinear scattering occurs. The wave changes the medium properties, and real or virtual emission of a wave by the particle is possible. This wave may now interact with

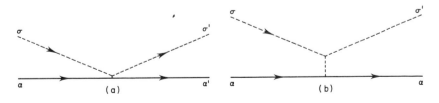

Fig. 5(a). Diagram illustrating the scattering of a wave by a particle. (b). Diagram illustrating the scattering of a wave by a particle emitting a real or virtual wave. This is absorbed by the incoming wave which then emits another wave.

an incoming wave (e.g. a t-wave) in a three-wave process: it may be absorbed by the incoming wave and another t-wave may be emitted. This is illustrated in Fig. 5(b) by the more complicated vertex. The emission of the wave by the particle may be a Cherenkov process if the disturbed medium has a phase velocity larger than the particle velocity. It can be of the undulator type if the particle carries out a suitable periodic motion in the disturbed medium.

The probability of such scattering processes is calculated in Tsytovich's book [3]. Compton and nonlinear scattering, when superposed may reinforce each other or partially cancel. In quantum language we say that this happens because the probability amplitudes with phase factors must be added and squared. Tsytovich shows that the ion scattering cross-section is much larger than that for electrons because in their case the nonlinear and Compton-scattering phase difference is nearly 180°. An important modification of Thomson scattering may take place if an ion acoustic or a Langmuir excitation is present in the plasma, e.g. as a thermal fluctuation. An incoming t-wave may interact with these longitudinal waves which are termed turbulent if their phases are random. The frequency shift $\Delta\omega$ of the scattered wave and the wave number shift of the scattered beam are again determined by the conservation laws

$$\Delta\omega = \omega_l$$
$$\Delta\mathbf{k} = \mathbf{k}_l \tag{63}$$

Such scattering is used to determine the level of plasma turbulence and we shall discuss this in the chapter on diagnostic methods. The scattering was theoretically investigated by Salpeter [6] and a review of theory and experiments by Evans and Katzenstein may be found in [7]. We indicate briefly the salient features of the result of the analysis.

The Thomson-scattered intensity $I(\omega)$ is proportional to the incoming intensity I_0, the Thomson cross-section σ_T, the electron density N_e and a factor $S(k, \omega)$, the dynamic form factor. Thus

$$I(\omega) = I_0 \sigma_T N_e S(k, \omega) \tag{64}$$

where g is a geometric factor and

$$S(k, \omega) = \frac{1}{\sqrt{\pi k N_e}} \exp\left(\frac{-(\omega - \omega_0)^2}{(k v_e)^2}\right), \qquad v_e = \left(\frac{2kT}{m_e}\right)^{1/2}. \tag{64a}$$

The Thomson cross-section σ_T is given by

$$\sigma_T = \frac{8\pi}{3} \pi e^2 / m_e c^2 = 6\cdot65 \,.\, 10^{-25} \text{ cm}^2.$$

It is so small that even for electron densities 10^{12}–10^{18} a MW laser is needed to give detectable scattered light. The factor (64a) is obtained when the electrons scatter independently. This happens when the laser wavelength λ is smaller than the Debye wavelength λ_D. When $\lambda \gg \lambda_D$ the Doppler broadening due to the electron thermal motion exhibited in formula (64a) is very much reduced leaving the total cross-section still equal to σ_T, and the dynamic form factor changes.

Indeed, an electron with thermal velocity v_e will cause scattered intensity to be Doppler-shifted in frequency by an amount of order kv_e and since $\lambda_D = (2kT_e/m)^{1/2}/\omega_p$, $kv_e = (2\pi\lambda_D/\lambda)\omega_p$. Thus when $\omega > \omega_p$ and $\lambda \gg \lambda_D$ the shift is small.

The spectrum of the scattered light can be characterized by the ratio $\alpha = (k\lambda_D)^{-1}$ and the dynamic form factor can be written in terms of it. It is given by the approximate expression (discussed in [7])

$$S(k, \omega)\, d\omega = \Gamma_\alpha(x_e)\, dx_e + Z\left(\frac{\alpha^2}{1 + \alpha^2}\right)^2 \Gamma_\beta(x_i)\, dx_i \qquad (65)$$

where

$$\beta^2 = Z\left(\frac{\alpha^2}{1 + \alpha^2}\right)\frac{T_e}{T_i}, \qquad \Gamma_\alpha(x) = \frac{\exp(-x^2)}{|1 + \alpha^2 W(x)|^2}, \qquad \Gamma_\beta(x) = \frac{\exp(-x^2)}{|1 + \beta^2 W(x)|^2}$$

$W(x)$ is a function of the variable $x = \Delta\omega/kv$, i.e.

$$\left.\begin{array}{c} x_e = \dfrac{\Delta\omega}{kv_e}, \qquad x_i = \dfrac{\Delta\omega}{kv_i} \\[3mm] \text{and} \\[2mm] W(x) = \left(1 - 2x\, e^{-x^2}\displaystyle\int_0^x e^{t^2}\, dt - i\sqrt{\pi}\, x\, e^{-x^2}\right), \end{array}\right\} \qquad (66)$$

and $\phi(x) = -2x^2 W(x)$ is the function (Chapter 4, equation (30b)) already introduced in the section on longitudinal waves.

The scattered intensity shows high peaks near the Thomson peak as indicated in Fig. 6 which shows the function $\Gamma_\alpha(x)$ for various parameter values. It is seen that, e.g. for $\alpha = 3, 4$, there are high peaks. These peaks arise at values of x where the real part of $1 + \alpha^2 W(x)$ or $1 + \beta^2 W(x)$ vanishes. Physically the resonances arise when plasma oscillations fulfil the wave–wave interaction conservation conditions. This is borne out by the lines in the upper part of Fig. 6 which indicate the values of x fulfilling the Bohm and Gross dispersion equation

$$x_0^2 = \tfrac{1}{2}(\alpha^2 + 3), \quad \text{i.e. } \omega_0^2 = \omega_p^2 + 3(kT/m)\, k^2.$$

For large values of λ/λ_D, the laser-wave sample regions of the plasma are larger than the Debye length, i.e. larger than the regions in which space neutrality is violated near an ion. Thus the ions are shielded by an electron cloud and there is a collective rather than an individual particle response. The

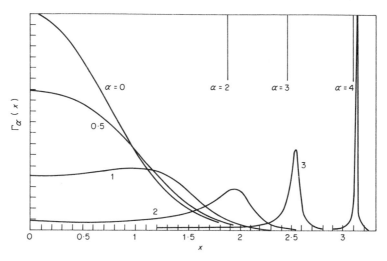

Fig. 6. The function $\Gamma_a(x)$ for various values of the parameter (indicated on curves). At the top, the position of the plasma frequency resonances given by the Bohm and Gross dispersion equation for the various values of α are shown. (From [6], with permission.)

observation direction selects the **k** value which the plasma oscillation adds to that of the laser according to (63) and at the frequency value corresponding to the solution of the dispersion equation the intensity peaks up.

SCATTERING OF PARTICLES BY WAVES

The scattering of particles by waves is a subject on which much has been written, which is not easy to summarize. It is important because it governs important transport properties like electrical and thermal conductivity. Taking a very general view, we can consider the conductivity to be determined by the time between collisions; the current density is given by

$$j = env \qquad (67)$$

and the velocity, in the case when a steady drift velocity is reached in an electric field E is given by

$$v = (eE/m)\tau \qquad (68)$$

where τ is the time between collisions. Classically, one calculates this time from gas-kinetic considerations of collisions with other particles but in a plasma in which plasma oscillations are excited, the electrons are also diverted from their straight paths because they are scattered by the waves and one can define an equivalent turbulent collision frequency or its inverse a turbulent collision time.

In fact drifting electrons create Langmuir turbulence as they are streaming against the stationary ions through the beam instability mechanism discussed above. The conductivity is given by

$$j/E = \sigma = \frac{e^2 n}{m}\tau = \varepsilon_0 \omega_p^2 \tau \tag{69}$$

and τ has been calculated by Buneman [8], on the basis of this mechanism, to be

$$\tau = 2\pi \omega_p^{-1}\left(\frac{M}{m}\right)^{1/3} \tag{70}$$

a value that is in agreement with some experiments [11]. Here the electrons lose energy by amplifying noise in the double stream instability frequency band to turbulence level. We shall return to this subject when absorption is discussed.

PARAMETRIC PROCESSES

So far, we have treated wave transformations statistically and assumed that the waves have random phases. This is appropriate when the spectral width of the radiation is larger than the growth rate, so that components of different frequency can get out of phase during the transformation process, i.e. when the phase change $\Delta\phi = 2\pi(v_1 - v_2)t$ in time $t = 1/\gamma$ is $\geqslant 2\pi$. When this is not the case a coherent treatment by means of coupled nonlinear equations is necessary. This gives different growth rates, because the transition probabilities calculated by Tsytovich are averaged over the phases. The coherent wave treatment was first given by Silin [21] and further developed by Goldman et al. [22].

Coherent wave transformation processes have a circuit analogue known in electric technology as the parametric amplifier or converter. Three L–C circuits are coupled by means of a nonlinear element, e.g. a capacity with a nonlinear amplitude response. It is driven by a "pump" frequency ω_0 and one of the circuits is tuned to it. Another is tuned to a frequency ω_1 and the third to the difference frequency. Owing to the nonlinear coupling, the pump excites an oscillation at frequency ω_1 or, when power at that frequency is also supplied, may amplify it. The circuit tuned to the frequency $\omega_0 - \omega_1$ is called the idler.

A mechanical analogy is that of two coupled pendula (illustrated in Fig. 7) coupled by an elastic link, the "parameter" with a spring constant s which varies with frequency as $s = s_0 + s_1 \cos \omega_0 t$. If $\omega_0 = \omega_1 + \omega_2$ where $\omega_1^2 = g_1/L_1$ and $\omega_2^2 = g_2/L_2$ it is possible for both pendula to become unstable. The coupler beats with oscillations at ω_2 to drive oscillations at ω_1 which in turn beat with ω_0 to drive oscillations at ω_2. The oscillation builds up the initial noise, or Brownian movement.

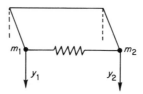

Fig. 7. Two pendula coupled by a spring with frequency-dependent restoring force.

In our case the incident laser light at frequency acts as the pump and its amplitude is $Z = 2Z_0 \cos \omega_0 t$. It may be coupled, e.g. to an ion acoustic wave X mode at frequency ω_2 and a Langmuir wave mode Y at frequency ω_1. We assume that in the absence of coupling

$$\omega_0 = \omega_1 + \omega_2 + \Delta \tag{71}$$

i.e. we allow a "mismatch" $\Delta \ll \omega_0$. The situation was analysed by Nishikawa [9] in a very instructive fashion. He starts with the coupled equations of a very general nature

$$\left.\begin{aligned}
\left\{\frac{d^2}{dt^2} + 2\Gamma_1 \frac{d}{dt} + \omega_1^2\right\} X(t) &= \lambda Y(t) Z(t) \\[2mm]
\left\{\frac{d^2}{dt^2} + 2\Gamma_2 \frac{d}{dt} + \omega_2^2\right\} Y(t) &= \mu Y(t) Z(t)
\end{aligned}\right\} \tag{72}$$

where λ and μ are coupling constants while Γ_1 and Γ_2 are damping constants of the uncoupled modes which may all be obtained from the macroscopic plasma equations applying to the particular problem in hand. It will be assumed that $\lambda\mu > 0$. Writing

$$X(t) = \int X(\omega)\, e^{-i\omega t}\, d\omega, \quad Y(t) = \int Y(\omega)\, e^{-i\omega t}\, d\omega$$

the Fourier transforms $X(\omega)$, $Y(\omega)$ are now related by the equations

$$\left.\begin{aligned}
[\omega^2 - \omega_1^2 + 2i\Gamma_1\omega]\, X(\omega) + \lambda Z_0[Y(\omega - \omega_0) + Y(\omega + \omega_0)] &= 0 \\[2mm]
[\omega^2 - \omega_2^2 + 2i\Gamma_2\omega]\, Y(\omega) + \mu Z_0[X(\omega - \omega_0) + X(\omega + \omega_0)] &= 0.
\end{aligned}\right\} \tag{73}$$

It is seen that $X(\omega)$ is coupled with $Y(\omega \pm \omega_0)$ which in turn are coupled with $X(\omega)$ and $X(\omega \pm 2\omega_0)$. We assume that the frequency-matching condition (71) is satisfied and also that $\mathrm{Re}(\omega) \approx \omega_1$. Neglecting $X(\omega + 2\omega_0)$ as being off resonance and treating the amplitude Z_0 of the pump signal Z as constant one obtains three equations in $X(\omega)$, $Y(\omega + \omega_0)$ and $Y(\omega - \omega_0)$. Setting the determinant of the coefficients equal to zero, one obtains

$$\det \begin{vmatrix} \omega^2 - \omega_1^2 + 2i\Gamma_1\omega & \lambda Z_0 & \lambda Z_0 \\ \mu Z_0 & A(+) & \\ \mu Z_0 & 0 & A(-) \end{vmatrix} = 0 \quad (74)$$

$$A(\pm) = (\omega \pm \omega_0)^2 + 2i\Gamma_2(\omega \pm \omega_0) - \omega^2.$$

which is the secular equation which determines the frequency and damping (or growing) of these modes.

Writing

$$\omega = \chi + i\gamma$$

then χ and γ respectively are the frequency and damping rate of the new mode and it will be unstable if $\gamma > 0$. γ depends on Z_0 and the value of this quantity, the pump field amplitude which makes γ just equal to zero is the threshold of the instability.

Various cases may be singled out by the analysis of the solutions of (74). If ω_1 is large compared with its frequency shift $Y(\omega + \omega_0)$ is off resonance and the equations simplify. The pump power is $K = \lambda\mu Z_0^2$.

A threshold

$$\lambda\mu Z_0^2 = K_c(\Delta) = K_m\{1 + \Delta^2/(\Gamma_1 + \Gamma_2)^2\} \quad (75a)$$

is obtained where

$$K_m = 4\omega_1\omega_2\Gamma_1\Gamma_2, \qquad \Delta = \omega_0 - \omega_1 - \omega_2 \quad (75b)$$

and a minimum threshold is obtained for exact matching, i.e. $\Delta = 0$ and in this case $K_c = K_m$.

At a threshold, the frequency of X is given by

$$\chi = \omega_1 + \Gamma_1\Delta/(\Gamma_1 + \Gamma_2) \quad (76)$$

and that of Y by

$$\chi - \omega_0 = -\omega_2 - \Gamma_2\Delta/(\Gamma_1 + \Gamma_2) \quad (77)$$

and it is seen that at the minimum threshold the frequency shifts vanish.

In the case of low frequency oscillations $|\omega_1| \ll \omega_2$, e.g. when X represents a long wave acoustic oscillation, the frequency shift may become comparable to ω_1 so that $Y(\omega + \omega_0)$ cannot be neglected. In this case the full equations allow solutions which are purely growing at zero frequency

$$\omega = i\gamma, \qquad \chi = 0$$

or

$$\omega = \chi + i\gamma, \qquad \chi \neq 0.$$

The instability resulting in the first case has been termed "oscillating two stream instability". We shall come back to it below.

If we put

$$\delta = \omega_0 - \omega_2 \qquad (78)$$

the threshold power can be written

$$K_c(\delta) = -\frac{\omega_1^2 \omega_2}{\delta}[\Gamma_2^2 + \delta^2] \qquad (79)$$

and thus turns out to be independent of Γ_1. We must have $\delta < 0$, i.e. $\omega_0 > \omega_2$ for this instability to occur.

When $\chi \neq 0$ the threshold is given by

$$K_c(\delta) = \frac{\Gamma_1 \Gamma_2}{\delta}\left\{4\delta^2 + \frac{[\Gamma_2^2 + 2\Gamma_1\Gamma_2 + \omega_1^2 - \delta^2]^2]}{(\Gamma_1 + \Gamma_2)^2}\right\} \qquad (80)$$

and it is seen that there is no threshold when the normal modes X and Y are both unattenuated, a feature also shown by formula (9). We now have $\delta > 0$.

For either type of instability the maximum growth rate well above threshold is found to be proportional to the cube root of the incident power.

GENERAL PROCEDURE FOR OBTAINING THE EQUATIONS OF COHERENTLY INTERACTING WAVES

The analysis of the preceding section treats the pump amplitude as constant in space, i.e. as having infinite wavelength. For laser light interactions we must treat finite wavelength and we shall now set up the appropriate equations. There are many 3-wave processes which can be described by equations derived by the general procedure to be outlined. They all have one feature in common: rapidly oscillating fields of different plasma modes interact on a much slower time scale, or even multiple time scales of increasing length. This is made very clear by analysis which proceeds from the outset to consider time scales and corresponding space scales of order $\varepsilon^{-1}, \varepsilon^{-2}, \ldots$, etc., where ε is a small quantity and by using [12]

$$\mathbf{x}_0, \mathbf{x}_1, \mathbf{x}_2, \ldots, \mathbf{x}_N; t_0, t_1, t_2, \ldots, t_c \qquad \text{when} \quad \mathbf{x}_n = \varepsilon^n \mathbf{x}, \quad t_n = \varepsilon^n t \qquad (81)$$

as independent space and time variables corresponding to the different space and time scales. The dependent variable may be the density, field or velocity, which we denote by $\mathbf{q}$ in general and we express it as

$$\mathbf{q}(\mathbf{x}_0, \ldots, \mathbf{x}_N; t_0, \ldots, t_N) = \sum_{n=1}^{N} \varepsilon^n \mathbf{q}_n(\mathbf{x}_0, \ldots, \mathbf{x}_N; t_0, \ldots, t_N) + O(\varepsilon^{N+1})$$

$$= \varepsilon \mathbf{q}_1, + \varepsilon^2 \mathbf{q}_2 + \varepsilon^3 \mathbf{q}_3 + \ldots. \tag{82}$$

We shall treat a set of coherent interacting waves for which q_1 will have the form

$$\mathbf{q}_1 = \text{Re} \sum_j \mathbf{a}_j^{\pm}(\mathbf{x}_0, \mathbf{x}_1; t_0, t_1) \exp[i(\mathbf{k}_j \cdot \mathbf{x}_0 \mp \omega_j t)] = \text{Re} \sum_j \mathbf{q}_{1j}. \tag{83}$$

We start with $n = 1$ because we deal with a perturbation. The subscript j assigns a number 1, 2 or 3 to each wave type, e.g. elementary, Langmuir or acoustic; the $a^{\pm}$ are the corresponding amplitudes of forward and backward moving waves, k_j and ω_j their unperturbed wave numbers and frequencies. In order to separate out phenomena happening at different space and time scales $\partial/\partial x$ and $\partial/\partial t$ are replaced by

$$\frac{\partial}{\partial x_0} + \varepsilon \frac{\partial}{\partial x_1} + \varepsilon^2 \frac{\partial}{\partial x_2} + \ldots; \qquad \frac{\partial}{\partial t_0} + \varepsilon \frac{\partial}{\partial t_1} + \varepsilon^2 \frac{\partial}{\partial t_2} + \ldots \tag{84}$$

accordingly the procedure is known as the derivative expansion technique. We apply it to an example where the differential equations are

$$\left(\frac{\partial^2}{\partial t^2} - C^2 \frac{\partial^2}{\partial x^2} + \Omega^2\right) \mathbf{q} = \frac{\partial}{\partial t} \lambda \mathbf{q}\mathbf{q} \tag{85}$$

but in order to present the general method at the same time we write the more general equation

$$\mathbf{P}\left(\frac{\partial}{\partial \mathbf{x}}, \frac{\partial}{\partial t}\right) \mathbf{q} = \mathbf{Q}\left(\frac{\partial}{\partial \mathbf{x}}, \frac{\partial}{\partial t}\right) \mathbf{q}\mathbf{q}$$

or

$$P_{ij}\left(\frac{\partial}{\partial \mathbf{x}}, \frac{\partial}{\partial t}\right) q_j = Q_{ijk}\left(\frac{\partial}{\partial \mathbf{x}}, \frac{\partial}{\partial t}\right) q_j q_k + O(\mathbf{q}^3) \left. \begin{array}{c} \\ \\ \\ \\ \\ \end{array} \right\} \tag{86}$$

in terms of the tensor operator functions, $\mathbf{P}$ and $\mathbf{Q}$, of the derivatives $\partial/\partial x$ and $\partial/\partial t$. In our special case

$$\mathbf{P}\left(\frac{\partial}{\partial \mathbf{x}}, \frac{\partial}{\partial t}\right) = \frac{\partial^2}{\partial t^2} - C^2 \frac{\partial^2}{\partial x^2} + \Omega^2, \qquad \mathbf{Q}\left(\frac{\partial}{\partial \mathbf{x}}, \frac{\partial}{\partial t}\right) = \lambda \frac{\partial}{\partial t}, \tag{87}$$

where $\lambda = \lambda_{ijk}$. The operators and the functions q on which they act are expanded and terms with the same power of ε are equated. The first two terms of zero and first order in ε in the expansions of the operators $\mathbf{P}$ and $\mathbf{Q}$ are given by

$$\mathbf{P}_0 = \frac{\partial^2}{\partial t_0^2} - C^2 \frac{\partial^2}{\partial x_0^2} + \Omega^2 \qquad (88a)$$

$$\mathbf{P}_1 = \varepsilon \left(2 \frac{\partial^2}{\partial x_0 \, \partial t_1} - 2C^2 \frac{\partial^2}{\partial x_0 \, \partial x_1} \right) \qquad (88b)$$

$$\mathbf{Q}_0 = \lambda \frac{\partial}{\partial t_0}, \qquad \mathbf{Q}_1 = \varepsilon\lambda \frac{\partial}{\partial t_1}. \qquad (88c)$$

Using these expansions the first and second order perturbations obey equations

$$\mathbf{P}_0 q_1 = 0 \qquad (89)$$

$$\mathbf{P}_0 q_{2i} = \frac{\partial}{\partial t_0} \lambda_{ikl} q_{1k} q_{1l} - \mathbf{P}_1 q_{1i} + 0(\varepsilon^3). \qquad (90)$$

Substituting (83) for q_1, we obtain

$$\mathbf{P}_0 q_1 = \sum_j (- \omega_j^2 + c^2 k_j^2 + \Omega^2)\, \mathbf{a}_j^{\pm} \exp[i(\mathbf{k}_j \cdot \mathbf{x}_0 \mp \omega_j t_0)] =$$

$$\sum D_j(\omega_j, \mathbf{k}_j)\, \mathbf{a}_j^{\pm} \exp[i(\mathbf{k}_j \cdot \mathbf{x}_0 \mp \omega_j t_0)] = 0 \qquad (91)$$

where $D_j(\omega_j, \mathbf{k}_j) = 0$ is the dispersion equation for the jth wave. From the right-hand side of equations (90) and (88b) we get

$$\mathbf{P}_0 q_2 = \mp \sum_l \sum_k i(\omega_l + \omega_k)\, \lambda \mathbf{a}_l^{\pm} \mathbf{a}_k^{\pm} \exp\{i[(\mathbf{k}_l + \mathbf{k}_k)\, \mathbf{x}_0 \mp (\omega_l + \omega_k)\, t_0]\}$$

$$\pm i \sum_j \left(\frac{\partial D_j}{\partial \omega_j} \frac{\partial}{\partial t_1} \mp \frac{\partial D_j}{\partial \mathbf{k}_j} \frac{\partial}{\partial \mathbf{x}_1} \right) \mathbf{a}_j^{\pm} \exp[i(\mathbf{k}_j \cdot \mathbf{x}_0 \mp \omega_j t_0)]. \qquad (92)$$

The last term is a solution of the homogeneous equation and can therefore not appear on the longer timescale. To avoid this secularity, as it is called in celestial mechanics, we must make this term cancel and enforce the condition $\mathbf{P}_0 q_2 = 0$ unless the $a_j^{\pm}$ are constant on this time scale. (See e.g. [4] for this procedure). This requires the matching conditions

$$\mathbf{k}_j = \mathbf{k}_l + \mathbf{k}_k; \qquad \omega_j = \omega_l + \omega_k. \qquad (92a)$$

In order to get the desired cancellation we must pick terms satisfying (92a) and obtain for forward waves the equations

$$\frac{d\mathbf{a}_1}{dt} = \left(\frac{\partial}{\partial t} + \mathbf{v}_{g1} \cdot \frac{\partial}{\partial \mathbf{x}}\right) \mathbf{a}_1(\mathbf{x}, t) = \lambda_1[(\omega_2 + \omega_3)\mathbf{a}_2\mathbf{a}_3] \Big/ \left(\frac{\partial D}{\partial \omega}\right)_{\omega_1, k_1} \tag{93a}$$

$$\frac{d\mathbf{a}_2}{dt} = \left(\frac{\partial}{\partial t} + \mathbf{v}_{g2} \cdot \frac{\partial}{\partial \mathbf{x}}\right) \mathbf{a}_2(\mathbf{x}, t) = \lambda_2[(m_1 - \omega_3)\mathbf{a}_1\mathbf{a}_3^*] \Big/ \left(\frac{\partial D}{\partial \omega}\right)_{\omega_2, k_2} \tag{93b}$$

$$\frac{d\mathbf{a}_3}{dt} = \left(\frac{\partial}{\partial t} + \mathbf{v}_{g3} \cdot \frac{\partial}{\partial \mathbf{x}}\right) \mathbf{a}_3(\mathbf{x}, t) = \lambda_3[(\omega_1 - \omega_2)\mathbf{a}_1\mathbf{a}_2^*] \Big/ \left(\frac{\partial D}{\partial \omega}\right)_{\omega_3, k_3} \tag{93c}$$

Here we have used the fact that the group velocity is given by (see p. 383 of [5] of Chapter 9)

$$\mathbf{v}_{gj} = \left(\frac{\partial D_j}{\partial \mathbf{k}_j}\right)_{\omega_j, \mathbf{k}_j} \Big/ \left(\frac{\partial D_j}{\partial \omega_j}\right)_{\omega_j, \mathbf{k}_j}, \tag{94}$$

and suppressed the suffix 1 of t and x indicating the space–time scale.

We can introduce damping constants and relax condition (92) by allowing a mismatch Δk in wave number and $\Delta \omega$ in frequency. We can also incorporate damping constants γ_i, $i = 1, 2, 3$, for the three waves and obtain equations of the type

$$\left(\frac{\partial}{\partial t} + \mathbf{v}_{g1} \cdot \frac{\partial}{\partial \mathbf{x}} + \gamma_1\right) \mathbf{a}_1(\mathbf{x}, t) = K_1 \mathbf{a}_2 \mathbf{a}_3 \exp[i(\Delta \mathbf{k} . \mathbf{x} - \Delta \omega t)] \tag{95a}$$

$$\left(\frac{\partial}{\partial t} + \mathbf{v}_{g2} \cdot \frac{\partial}{\partial \mathbf{x}} + \gamma_2\right) \mathbf{a}_2(\mathbf{x}, t) = K_1 \mathbf{a}_1 \mathbf{a}_3^* \exp[i(\Delta \mathbf{k} . \mathbf{x} - \Delta \omega t)] \tag{95b}$$

$$\left(\frac{\partial}{\partial t} + \mathbf{v}_{g3} \cdot \frac{\partial}{\partial \mathbf{x}} + \gamma_3\right) \mathbf{a}_3(\mathbf{x}, t) = K_3 \mathbf{a}_1 \mathbf{a}_2^* \exp[i(\Delta \mathbf{k} . \mathbf{x} + \Delta \omega t)] \tag{95c}$$

where the K's are coupling coefficients and

$$\Delta \mathbf{k} = \mathbf{k}_1 - \mathbf{k}_2 - \mathbf{k}_3, \qquad \Delta \omega = \omega_1 - \omega_2 - \omega_3. \tag{96}$$

A general solution of the three equations can be found in terms of elliptic functions as we shall see below. They simplify when, as in the special case of Nishikawa's analysis one wave, say a_1, the pump wave, has constant amplitude and the space dependence is ignored. Alternatively, the amplitudes may be constant in time so that coupling occurs in space. One then finds a dispersion equation of the form

$$(k - i\alpha_2)\left(k + \Delta k - i\frac{\gamma_3}{v_{g3}}\right) + \frac{K_2 K_3}{v_{g2} v_{g3}} |a_1|^2 = 0 \tag{97}$$

and spatial growth is obtained with Im $k = \alpha$ given by

$$\alpha^2 = K_2 K_3 |a_1|^2 / v_{g2} v_{g3} \tag{98}$$

when $\Delta k = \gamma_3 = 0$. Thus when $v_{g2} v_{g3} > 0$ there will be exponential growth in space and when $v_{g2} v_{g3} < 0$ there will be periodic backward amplification or oscillation.

Equations (93, 95) are, of course, similar in structure to equation (23) previously derived for the incoherent case, from quantal considerations. We have postulated the form (86) which must be established from the basic equations used for any particular problem. If the model of the macroscopic continuity momentum and Maxwell equation is adopted, equations of this form are found for the $t \to t + s$, $t_0 \to t_1 + t_2$, $t \to t + l$, process. For processes involving l and s with Landau damping the Vlasov equation is appropriate.

The interactions take place when the dispersion equations are simultaneously satisfied for the component waves. This is analogous to the crossing of dispersion curves discussed in connection with the discussion of convective and absolute instabilities. The nature of the instability can again be established. This has been done in many cases, e.g. by Rosenbluth *et al.* [17].

A particular phenomenon, the explosive instability, is best discussed in connection with the sign of the wave energy. It may be shown that the wave energy is given by

$$\tfrac{1}{4}\omega \left(\frac{\partial D}{\partial \omega} \right)_{\omega, k} |E_{\omega k}|^2. \tag{99}$$

This follows from expression (26) which we have already used and the derivation may be found in detail in the book by Wieland and Wilhelmsson [5]. According to the sign of $(\partial D / \partial \omega)_{\omega, k}$ this energy may be positive or negative. It should be pointed out that this is an r-f energy of perturbation relative to an unperturbed state. In setting up a wave of negative energy the rest of the system gains energy instead of losing energy as it does in the more familiar case when a positive energy wave is set up. On the other hand absorption of a negative energy wave costs energy.

Chu [15], Sturrock and others pointed out at an early stage that the concept of negative energy waves is helpful in connection with beam–wave interaction and showed that instability occurs as a result of interaction between waves of opposite sign.

If one of the three denominators in equations (93) associated with the highest frequency is negative, an instability results which grows to infinite amplitude in a finite time t_∞. The growth near $t = t_\infty$ behaves like $(t - t_\infty)^{-2}$ and the instability is termed explosive. This subject is discussed in [5, 10]. Here we add, that the phenomenon may be understood with reference to the Manley–Rowe relations already discussed above, which can again be derived from the present set of equations (95) by multiplying with a_1^*, a_2^*, a_3^* and adding the conjugates. The negative energy sign has the effect of leading to Manley–Rowe type relations

$$\frac{dA_1}{dt} = \frac{dA_2}{dt} = \frac{dA_3}{dt} \tag{100a}$$

where for suitable choice of the a_i's,

$$\lambda_i a_i a_i^*(\partial D/\partial \omega)/\omega_i = A_i \tag{100b}$$

are the respective actions and thus all amplitudes may grow simultaneously.

Having discussed the general features of coherent interaction we shall consider the case of the parametric Langmuir-ion acoustic interaction $t \to l + s$ in more detail. Since ω_s is very small compared to ω_T this occurs when $\omega_T \approx \omega_p$. Hence $\mathbf{k}_T$ is very small and for momentum conservation we must have $\mathbf{k}_l + \mathbf{k}_s = 0$ and $\mathbf{k}_l$ and $\mathbf{k}_s$ are almost perpendicular to $\mathbf{k}_T$. This is illustrated in Fig. 8. We first assume that t and s interact resonantly. But we also

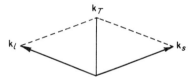

Fig. 8. k-Vector diagram for transverse wave decaying into an ion acoustic and a Langmuir wave.

consider the case when there is a frequency mismatch of the order of ω_s and in this case we can also include an off-resonant ion acoustic wave. Figure 9 shows the vector diagrams for two 3-wave processes which are coupled. The ion acoustic wave of the right-hand diagram is included in the left-hand one as an off-resonant wave. The left-hand ion acoustic wave has the same frequency as the right-hand one but is coupled off-resonantly to the (different) Langmuir wave of the right-hand diagram.

The transverse wave is described by its electric field $\mathbf{E}_T(x, t) = \hat{e}_T E$

$$\mathbf{E}_T(x, t) = \text{Re}\{\hat{e}_T E_0^+(x, t) \exp[i(\mathbf{k}_T \cdot \mathbf{x} + \omega_T t)]\} \tag{101a}$$

and the Langmuir wave and the ion wave by their densities n_{e1}^l and n_{l1}^s

$$n_{e1}^l(\mathbf{x}, t) = \text{Re}\{N_{l1}(\mathbf{x}, t) \exp[i(\mathbf{k}_{e1}\mathbf{x} - \omega_{e1}t)]\} \tag{101b}$$

$$n_{e1}^s(\mathbf{x}, t) = \text{Re}\{N_{s1}^\pm(\mathbf{x}, t) \exp[i(\mathbf{k}_s \cdot \mathbf{x} \pm \omega_s t)]\}. \tag{101c}$$

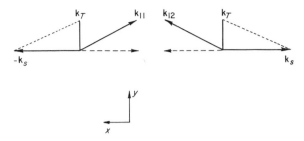

Fig. 9. k-Vector diagram illustrating coupling of two three-wave processes.

Where E_0^+, N_l, $N_s^\pm$ vary on the slow time scale and $\hat{e}_T$ is a unit vector. By means of the derivative expansion technique we obtain in the first case

$$\left(\frac{\partial}{\partial t} + v_g \frac{\partial}{\partial \mathbf{x}} + \gamma_T\right) E_T(\mathbf{x}, t) = - C_{es} N_s^+ N_{e1} \, e^{i\phi t} \tag{102}$$

with coupling coefficient C_{es}

$$C_{es} = \frac{e\omega_{e1}}{4\pi n_0 \varepsilon_0 k_{e1}} (\mathbf{k}_{e1} \cdot \hat{e}_T) \tag{102a}$$

and a small mismatch $\phi = \omega_T - \omega_e - \omega_s$

$$\gamma_T = \omega_{pe} v_e / 2\omega_T^2, \qquad v_{gT} = \frac{c^2 k_T}{\omega_T}. \tag{102b}$$

The equations for N_l and N_s^+ are

$$\left(\frac{\partial}{\partial t} + \mathbf{v}_{gl_1} \cdot \frac{\partial}{\partial \mathbf{x}} + \gamma_T\right) N_l = - C_{sT} E_T N_s^{+\,*} \, e^{i\phi t} \tag{102c}$$

$$\left(\frac{\partial}{\partial t} + \mathbf{c}_s \cdot \frac{\partial}{\partial \mathbf{x}} + \gamma_s\right) N_s^+ = C_{Tl} E_T N_l^* \, e^{i\phi t} \tag{102d}$$

with coupling coefficients

$$C_{sT} = \frac{ek_{l1}}{4m_e \omega_l} (\mathbf{k}_{l1} \cdot \hat{e}_T), \qquad C_{Tl} = \frac{e\omega_e k_s}{4m_i \omega_{T^{(i)}}} (\mathbf{k}_s \cdot \hat{e}_T) \tag{102e}$$

and with $v_{gl} = (|k_l|\gamma_l/\omega_e) v_{Te}^2$ and damping constants $\gamma_l = v_e/2$, $\gamma_s = v_i/2$. It is convenient to introduce new variables

$$\alpha_e = N_l, \qquad \alpha_s = N_s^* \, e^{-i\phi t}$$

and to assume that these amplitudes, α_e, α_s, vary as $\exp[i(qx - \Omega t)]$. When the pump wave is constant ($E_T = \text{constant}$) one easily obtains the dispersion equation

$$(\Omega - \mathbf{q}.\mathbf{v}_{gl} + i\gamma_l)(\Omega - \mathbf{c}_s.\mathbf{q} - \phi + i\gamma_s) + C_{sT}C_{Tl}|E_T|^2 = 0. \quad (103)$$

In the Nishikawa case, $q = 0$, unstable solutions for $\Omega = x + iy$ are obtained, the threshold condition for growing waves ($\gamma > 0$) becomes

$$C_{sT}C_{Tl}|E_T|^2 = \gamma_s\gamma_l + \frac{\phi^2\gamma_s\gamma_l}{(\gamma_s + \gamma_l)^2} \quad (104a)$$

the growth rate is

$$y = \frac{1}{2}\left(\frac{e^2k_s^2}{4m_em_i\omega_l\omega_s}|E_T|^2(\mathbf{k}_l.\hat{e}_T)\right)^{1/2}. \quad (104b)$$

In the case of the coupled diagrams of Fig. 9 one obtains equations for E_T, N_{e1}, N_{e2}, N_s^+, N_s^-, which we shall not write down in detail. One now obtains a generalization of Nishikawa's dispersion equation for propagation in the x-direction (Fig. 9)

$$(\Omega - qv_{gl1x} + \delta + i\gamma_l)(\Omega - qv_{gl2x} - \delta + i\gamma_l)(\Omega + \omega_s + i\gamma_s)$$
$$\times (\Omega - \omega_s + i\gamma_s) + K\delta\omega_s = 0 \quad (105a)$$

where coupling constants

$$K = -4C_{sT}C_{Tl}|E_0|^2 \quad (105b)$$

and the x-components of the group velocities v_{gl1x}, v_{gl2x} figure and where

$$\Omega = \omega - \delta, \qquad \delta = \omega_T - \omega_l, \quad (105c)$$

and amplitudes are assumed to vary as

$$\exp[i(qx - \omega t)].$$

We shall not discuss it in detail but draw attention to the important result obtained when $q = 0$, namely that two solutions are possible

(i) $\Omega = iy \quad (x = 0)$†

(ii) $\Omega = x + iy \quad (x \neq 0)$, $\qquad y + \gamma_s = (y + \gamma_l)/F(x, y)$

$$F(x, y) = \frac{1}{K\omega_s\delta}\{(\delta + x)^2 + (y + \gamma_l)^2\}\{(\delta - x)^2 + (y + \gamma_l)^2\}$$

Solution (i) corresponds to the case where the ion acoustic modes have undergone frequency shifts so that they are purely growing (non-propagating, Re $\Omega = 0$) density perturbations with the two Langmuir waves shifted to the

† These x, y should not be confused with the space directions of Fig. 9.

frequency of the transverse wave. This instability is commonly called the oscillating two-stream instability (OTS) for no particular good reason other than an analogy to the two-stream instability. The streaming motion of the OTS is caused by the driving field but the electron distribution is double-humped in both cases.

We have already obtained this result from Nishikawa's formal analysis. Now we see that because of the large frequency shift, it is really a five-wave process. We shall obtain further physical insight in terms of the action of the ponderomotive force discussed below.

Solution (ii) corresponds to the parametric decay instability with the transverse wave decaying into a Langmuir and an ion sound wave.

PROCESSES OF INTEREST FOR LASER FUSION

Two processes important for laser fusion have been named in analogy with molecular physics terminology. Raman scattering arises when light is scattered from molecular vibrations, thought to be analogous to plasma oscillations and therefore the process $t \to t + l$ is called a Raman process.

Brillouin scattering involves light scattering by a sound wave and therefore the process $t \to t' + s$ is called a Brillouin process. The analogy is not perfect. We have seen that the pump wave stimulates growth of the Langmuir, or ion sound waves respectively and therefore one speaks of stimulated Raman and Brillouin scattering.

Equations of type (93) may be established for these processes and the dispersion equations are obtained and from them the growth rates and thresholds, much as in the special case analysed by Nishikawa. The results together with detailed references are listed in [10], [23].

It is important to note that Raman and Brillouin processes may lead to backscattering as indicated in Figs 10 and 11. Another process of importance is the decay, $t \to t + l$, of a transverse wave into two Langmuir waves, the two-plasmon decay. It happens when $\omega_0 = 2\omega_p$, i.e. when the plasma density is given by $N = \omega_l^2 \varepsilon_0 m/e^2$ which is one quarter of the critical density. The growth rate γ of the two-plasmon decay is given by (e.g. [10])

$$\gamma^2 = \frac{k_0^2}{32} \left(\frac{eE}{m\omega_0} \right)^2.$$

The plasmon may then recombine with the t-wave to give a t-wave of frequency $\frac{3}{2}\omega$.

In assessing the importance of these instabilities for laser fusion, we must remember that the absorption region is not uniform, but shows a steep density gradient as in Fig. 12.

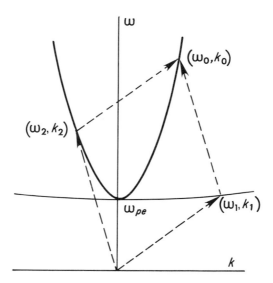

Fig. 10. k-Vector diagram illustrating backscattering in the case of a Raman process: the vector $\mathbf{k}_2$, has a component in the backward direction.

We have learned that light cannot penetrate into the over-dense region, and is either absorbed or reflected in a plasma layer near the region where the plasma frequency equals the laser frequency ω, i.e. where $n = m\omega^2\varepsilon_0/e^2$. Instabilities, however, lead to ion sound or plasma oscillation build-up at the expense of laser energy. The ion sound or plasma oscillations in turn are either Landau damped, or transformed by further nonlinear processes

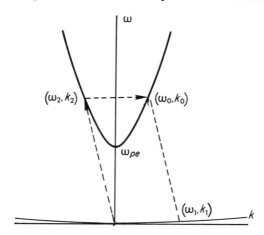

Fig. 11. k-Vector diagram illustrating backscattering in the case of a Brillouin process. The vector $\mathbf{k}_2$, has a component in the backward direction.

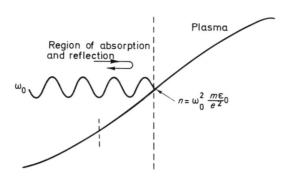

Fig. 12. Wave incident on an inhomogeneous plasma.

into random particle motion, i.e. heat. It is now thought that the conversion of plasma oscillations into random motion involves the formation of solitons and cavitons and their subsequent collapse, a phenomenon which we shall discuss below.

It would be most important to be able to predict the amount of laser energy reflected and the amount absorbed due to

(i) linear absorption;
(ii) nonlinear absorption, involving instabilities.

Unfortunately this task has not been accomplished to date. There are several formidable difficulties. One is that it is not enough to know the growth rates of instabilities. Growth does not continue indefinitely but is limited by further higher order nonlinear processes coming into play. These are, at present, not sufficiently understood. We shall discuss some of these processes and their nonlinear evolution below.

Another difficulty is the effect of inhomogeneity on growth rates. We have so far analysed processes in a uniform plasma but the effects of inhomogeneity are very important, and we shall now examine them.

In the case of the OTS, Perkins and Flick [16] have shown that the threshold is raised because the plasma wave convects out of the interaction region of length x_0 in a time

$$\tau = \int_0^{x_0} dx/v_g.$$

In the threshold expression, Γ is now replaced by $\Gamma + \tau^{-1}$. They solved the equations by WKB methods. In a more simple-minded fashion one can estimate the interaction region thickness by an estimate of the distance the wave has to travel until the propagation direction of the plasma wave becomes perpendicular to the E-field whereupon interaction stops. The inhomogeneous threshold I_i in a plasma with length scale $L = (1/n \, dn/dx)^{-1}$ of inhomogeneity is given in terms of the homogeneous one I_h by

$$I_i/I_h = \left(\frac{\Gamma_e}{\omega_p} + \frac{1}{2Lk_z}\right)$$

and the corresponding ratio for the $t \rightarrow l + s$ decay becomes

$$I_i/I_h = \left(\frac{\Gamma_e}{\omega_{pe}} + \frac{A}{k_z L}\right)$$

where A is the exponent of the amplification e^A of the Langmuir wave which has occurred in the unstable region. The problem has been analysed by Rosenbluth and co-workers and a review may be found in [17]. Their WKB analysis for the parametric problem starts with the two equations (see Chapter 7, [7], for this method of analysis)

$$\frac{\partial a_1}{\partial t} + \Gamma_1 a_1 + V_{g1} \frac{\partial a_1}{\partial x} = \gamma_0 a_2 \exp\left(i \int_0^x K \, dx'\right)$$

$$\frac{\partial a_2}{\partial t} + \Gamma_2 a_2 + V_{g2} \frac{\partial a_2}{\partial x} = \gamma_0 a_1 \exp\left(-i \int_0^x K \, dx'\right)$$

one gets (when the pump amplitude is constant) in a weakly inhomogeneous medium such that the WKB method is applicable. These equations are Laplace-transformed in time and the solutions are discussed by means of turning point techniques. The problem of Brillouin and Raman scattering is also analysed and the results are summed up in Table I .

They also show, that a necessary condition for the existence of an absolute instability is that not only the matching conditions $\Sigma k_i = 0$ holds but also $d/dz \, \Sigma k_i = 0$. It is interesting to note that the sophisticated analysis confirms results of more simple-minded considerations based on the time during which an instability convects out of the critical region [12]. Brillouin scattering is also investigated in an expanding plasma with a velocity u which is non-uniform with a scale length L_u. The ratio L_n/L_u is the density and velocity scale length ratio. The last column contains the author's conclusions about the nature of the instability.

These conclusions are important because absolute instabilities will tend to build up at the location of their generation, whereas convective instabilities move away and drift out of the interaction region. On the other hand if the convection is in the direction perpendicular to that of the incoming laser radiation, the product of the instability, e.g. the Langmuir waves still pile up.

The build-up time is extremely short. In the case of Brillouin backscatter, with a laser power of $10^{16} \, W \, m^{-2}$ it is of the order of tenths of picoseconds.

Table I. Threshold and growth rates for Raman and Brillouin scattering in an inhomogeneous plasma.

Instability	Plasma density	Growth rate	Threshold	Nature
Raman backscatter	$\omega_0 > 2\omega_p > \left(\dfrac{v_c}{c}\right)^2 \omega_0$	$\left(\dfrac{v_0}{c}\right)(\omega_0\omega_p)^{1/3}$	$(v_0/c)^2 k_0 L > 1$	Convective in x
Raman backscatter	$\omega_0 \sim 2\omega_p$	$\left(\dfrac{v_0}{c}\right)\omega_0$	$(v_0/c)^2 (k_0 L)^{4/3} > 1$	Absolute
Raman sidescatter	$\omega_0 > 2\omega_p > \left(\dfrac{v_c}{c}\right)\omega_0$	$\left(\dfrac{v_0}{c}\right)\omega_p$	$(v_0/c)^2 (k_0 L)^{4\ 3} > 1$ and $(v_0/c)(\omega_p/\omega_0)\, k_0 L_u > 1$	Temporally growing but convective in y
Brillouin backscatter	$\omega_p > \dfrac{T_i}{mc^2}\,\omega_0$	$\dfrac{v_0}{(2cc_g)^{1/2}}\,\omega_{pi}$	$\left(\dfrac{v_0}{c_c}\right)^2 \left(\dfrac{\omega_p}{\omega_0}\right) k_0 L_u > 1$	Convective in x
Brillouin sidescatter	$\omega_p > \dfrac{T_i}{mc^2}\,\omega_0$	$\left(\dfrac{v_0}{c}\right)\dfrac{\omega_0}{\omega_p}\left(\dfrac{L_n}{L_u}\right)^{1/2}\omega_{pi}$	$\left(\dfrac{v_0}{v_c}\right)^2 \left(\dfrac{L_n}{L_u}\right)(k_0 L_u)^{4/3}$ $(1+L_u/L_n)^{2/3} > 1$	Temporally growing but convective in y for finite k
Brillouin scatter (quasimode)	$\omega_p > \dfrac{T_i}{mc^2}\,\omega_0$	$\dfrac{3^{1/2}}{2^{1/3}}\,\omega_{pi}^{2/3}(k_0 v_0)^{1/3}$	$\left(\dfrac{v_0}{c}\right)^2 > (k_0\lambda_D)^3(\omega_{pi}\omega_0)$ and $\left(\dfrac{v_0}{v_c}\right)^2 (k_0 L_n) > 1$	Convective in x and y

Absolute instabilities can therefore build up large amplitudes of plasma or elm. waves which, in turn have important effects on the subsequent development of the plasma density distribution. They will be examined in the section devoted to the ponderomotive force.

While the effects of inhomogeneity of the light absorbing layer are important, particularly in the absorption region near the critical density layer, the interaction time, or length may allow the effects of nonlinear interaction, i.e. pump depletion to come into play. We shall therefore present the theory on nonlinear interaction which shows that the growth rates calculated before on the assumption that the attenuation of the incoming laser light is unimportant only hold in the initial phase; a saturation phenomenon sets in which in the case of a uniform plasma will tend to make the behaviour periodic and not one of exponential growth. Moreover, in the underdense region, two more instabilities can linearly or nonlinearly develop, the modulational and the filamentation instability, which will also be discussed below.

NONLINEAR BEHAVIOUR OF THREE-WAVE INTERACTION

The previous treatment of three-wave interaction has been linearized by assuming that the amplitude of the pump wave is constant. In fact, the growth of the instabilities cannot be exponential for all times, but must saturate. It is possible to obtain exact nonlinear solutions by allowing the pump wave amplitude to change, and to exhibit this effect. We shall sketch the derivation of these results and write the equations for the three-wave interaction in terms of action variables

$$A_n = \frac{1}{2} \frac{|\varepsilon_0 E_n|^2}{\omega_n} \qquad n = 0, 1, 2. \tag{106}$$

In terms of these variables (93) become

$$\left(\frac{\partial}{\partial t} + \mathbf{v}_0 \cdot \frac{\partial}{\partial \mathbf{x}} \right) A_0 = -i\Gamma A_1 A_2 \tag{106a}$$

$$\left(\frac{\partial}{\partial t} + \mathbf{v}_1 \cdot \frac{\partial}{\partial \mathbf{x}} \right) A_1 = -i\Gamma_1 A_0 A_2^* \tag{106b}$$

$$\left(\frac{\partial}{\partial t} + \mathbf{v}_2 \cdot \frac{\partial}{\partial \mathbf{x}} \right) A_2 = -i\Gamma_2 A_0 A_1^*. \tag{106c}$$

In the case of stimulated Brillouin backscattering

$$\Gamma = \frac{e^2}{4m_e \varepsilon_0} \left(\frac{\omega_s}{\omega_0 \omega_1} \right)^{1/2} \left(\frac{2n_e}{kT_e} \right)^{1/2}. \tag{106d}$$

On the left-hand side the group velocity appears, and not the phase velocity. This means that in the absence of coupling, linear analysis applies with amplitudes varying as a function of $x - v_n t$ only. This simple time dependence is lost when the waves interact. We treat the case when the interaction time is comparable to the pulse duration. Spatial derivatives can no longer be neglected and we look for the solution in a moving frame with coordinate $\xi = x - ut$. Representing the complex amplitudes in terms of real quantities $a_n(\xi)$ and $\phi_n(\xi)$ by

$$A_n = a_n(\xi) \exp[i\phi_n(\xi)]$$

and introducing

$$D_n = \Gamma/(v_n - u), \qquad \theta(\xi) = \phi_0(\xi) - \phi_1(\xi) - \phi_2(\xi) \qquad (107)$$

the resulting equations

$$\frac{\partial a_i}{\partial \xi} = -D_i n_j a_k \sin \theta(\xi), \qquad \frac{\partial \phi_i}{\partial \xi} = -D_i \frac{a_j a_k}{a_i} \cos \theta(\xi) \qquad (108)$$

where i, j, k are cyclic permutations of 0, 1, 2 lead to an integral relation

$$a_0 a_1 a_2 \cos \theta(\xi) = \Lambda \qquad (109)$$

where Λ has to be determined by initial or boundary conditions. The Manley–Rowe relations can again be obtained from (108) and they can be integrated to yield

$$D_1 a_0^2(\xi) + D_0 a_1^2(\xi) = D_1 a_0^2(0) + D_0 a_1^2(0) \qquad (110a)$$

$$D_2 a_0^2(\xi) + D_0 a_2^2(\xi) = D_2 a_0^2(0) + D_0 a_2^2(0) \qquad (110b)$$

$$D_1 a_2^2(\xi) - D_2 a_1^2(\xi) = D_1 a_2^2(0) - D_2 a_1^2(0). \qquad (110c)$$

We now look for solutions which are stationary in the frame moving with velocity u, i.e. they propagate without change of form. They are obtained in terms of Jacobi elliptic functions [18] $s_n(\mu, k)$, $c_n(\mu, k)$ and $d_n(\mu, k)$ which obey the relation

$$s_n^2(\mu, k) = 1 - c_n^2(\mu, k), \qquad d_n^2(\mu, k) = 1 - k^2 s_n^2(\mu, k). \qquad (111)$$

It is well known that they are periodic as functions of Re μ and, with another period, of Im μ. We sketch the procedure for obtaining this solution as follows: from (109) sin θ may be calculated and eliminated by means of the first equation (108) with $i = 1$. Now $a_1^2(\xi)$ may be expressed in terms of $a_0^2(\xi)$ and $a_2^2(\xi)$ by means of the conservation relations (110). The result is

$$\frac{\partial a_1^2}{\partial \xi} = 2D_1[a_0^2 a_1^2 a_2^2 - \Lambda^2]^{1/2}. \qquad (112)$$

We are free to choose the origin of the coordinate system so that a_1 has a maximum value. To get a solution of interest we take $a_0^2(0) = 0$ to achieve this and put $\Lambda = 0$. (112) now becomes

$$\frac{\partial a_1}{\partial \xi} = D_1 a_0 a_2. \tag{113}$$

Using the conservation relations (110) we get

$$\xi = [D_0 a_1^2(0) [D_1 a_2^2(0) - D_2 a_0^2(0)]^{-1/2} \tag{114}$$

$$\int_{a_1^2(0)}^{a_1^2(\xi)} \frac{da_1}{\left\{ \left(1 - \frac{a_1^2}{a_1^2(0)}\right) \left(1 + \frac{D_2 a_1^2(0)}{[D_1 a_2^2(0) - D_2 a_1^2(0)]} \frac{a_1^2}{a_1^2(0)}\right) \right\}^{1/2}}$$

and in terms of the variable $s^2 = 1 - a_1^2/a_1^2(0)$,

$$(D_1 D_0)^{1/2} a_2(0)\xi = \int_0^{(1 - (a_1^2(\xi)/a_1^2(0))^{1/2}} \frac{ds}{(1 - s^2)(1 - k^2 s^2)} \tag{115}$$

and so in view of the definition of the Jacobi elliptic functions in terms of the Jacobi integral

$$a_1^2(\xi) = a_1^2(0)c_n^2[(D_1 D_0)^{1/2} a_2(0)\xi, k] \tag{116}$$

where

$$k^2 = \frac{D_2 a_1^2(0)}{D_1 a_2^2(0)}. \tag{117}$$

Further on, by means of the conservation equations it is found that

$$a_0^2 = \frac{D_0}{D_1} a_1^2(0)s_n^2[(D_1 D_0)^{1/2} a_2(0)\xi, k] \tag{118}$$

$$a_1^2 = a_1^2(0)c_n^2[(D_1 D_0)^{1/2} a_0(0)\xi, k] \tag{119}$$

$$a_2^2 = a_2^2(0) d_n^2[(D_1 D_0^-)^{1/2} a_2(0)\xi, k] \tag{120}$$

In the case $u = 0$, Fig. 13 shows the oscillatory behaviour of the solutions. The pump wave is seen to vary periodically, decreasing each time until all its energy has been transferred to $a_1(\xi)$ and $a_2(\xi)$ [12]. When $u \neq 0$ we have an infinite wave-train modulated by the oscillations of the elliptic functions. In the special case, $k = 1$,

$$c_n(\mu, 1) = d_n(\mu, 1) = \text{sech } \mu, \quad s_n(\mu, 1) = \tanh \mu. \tag{121}$$

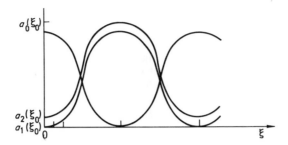

Fig. 13. Normalized action amplitudes a_0, a_1, a_2 of three non-linearly interacting waves.

The solutions are now:

$$a_0^2(\xi) = \frac{D_0}{D_1} a_1^2(0) \tanh^2(\beta\xi) \tag{122}$$

$$a_1^2(\xi) = a_1^2(0) \operatorname{sech}^2(\beta\xi) \tag{123}$$

$$a_2^2(\xi) = a_2^2(0) \operatorname{sech}^2(\beta\xi) \tag{124}$$

$$\beta = (D_1 D_3)^{1/2} a_2(0).$$

The solutions are pulses (waves with infinite period). We call them solitary waves and distinguish them from solitons which have yet another property which we cannot yet prove for our solutions, the property to reemerge with the same shape after collision with another soliton which we shall discuss below. (See Fig. 14 below for the shapes of these functions).

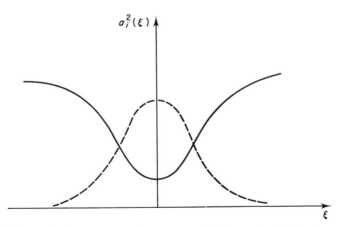

Fig. 14. Shapes of solitary waves. tanh x + constant (full curve), sech x (broken curve).

MODULATION AND FILAMENTATION INSTABILITIES

The OTS instability is an example of a purely growing instability. There are other instabilities of this kind, which are of importance in laser fusion.

Self-modulation occurs when the interaction of an electromagnetic wave with the plasma results in a modulation of the amplitude of the incident wave in the direction of propagation. This type of process was first described by Volkov [20] and more recently a number of authors have analysed self action due to various nonlinearities. Physically they result from the action of the ponderomotive force which we have discussed in the case of the OTS, from which it is indistinguishable in the case of Langmuir wave interaction when the pump wave has infinite wavelength [12]. Some remarkable properties of the modulational instability for Langmuir waves were discovered by Zakharov (Chapter 6 [11]) who showed that solitary waves emerge. Their subsequent collapse has been studied by several authors. We shall come back to this topic below.

The self modulation of an electromagnetic wave may be understood with reference to Fig. 15 as a four-wave interaction. (a) illustrates a k-vector diagram for $t \rightarrow t' + s$ proceeding in one dimension in the case of a standing wave pump wave. In the case of a travelling pump wave its components are shown as TO^+ and TO^- in Ω. Each of these components decays into a

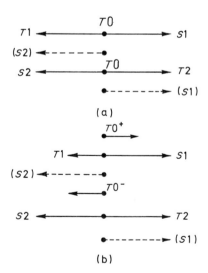

Fig. 15(a). k-Vector diagram for self modulation. Two three-wave processes are coupled by a sound wave with $k_s \| k_0$ in the case of a standing pump wave TO.

(b) k-Vector diagram for self modulation. Two three-wave interactions are coupled by a sound wave with $k_s \| k_0$ in the case of a travelling pump wave with components TO^+ and TO^-.

reflected electromagnetic wave and a sound wave of low frequency. As pointed out earlier, S2 must be considered in the decay of TO^+, and S1 in the decay of TO^- because the sound frequency is so low that the process is still resonant. It couples the two processes and as a result the four-wave interaction lowers the frequency of the wave TS1 as in the case of the OTS for $\delta < 0$. The same happens to TS2 so that both oscillations occur at the pump frequency ω_0. The k-vectors, i.e. the spatial period of these waves are, however, different from k_0 so that the two-wave patterns (which are purely growing waves in space) set up a beat pattern, the modulation. A similar phenomenon occurs when the transverse direction comes into play. This is illustrated in Fig. 16 which shows sound waves emitted in two opposite directions coupling two decays in the same manner. The terminology, Stokes and anti-Stokes wave, is used in analogy with that used in atomic physics, for the waves with up-shifted and down-shifted k's respectively.

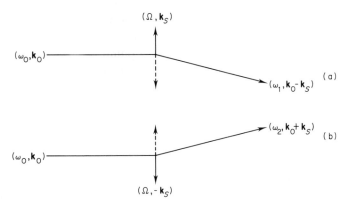

Fig. 16. k-Vector diagram for filamentation. Two three-wave interactions are coupled by a sound wave with $k_s \perp k_0$.

Again, a purely growing instability occurs, which periodically pumps wave energy into the transverse direction. It will therefore be concentrated into a number of long filaments (the interaction also sets up sound waves in a direction perpendicular to the plane of the figure) with a spacing of threads given by the wavelength of the growing sound wave.

The analytic treatment of four-wave interactions leads now to four coupled equations, analogous to the three treated above. The minimum threshold for the modulational instability

$$|E_{t_0}|^2 = 4\frac{m_e}{e^2}\gamma_e kT_e\omega_1 v_e \tag{125}$$

is the same as that for the OTS. Well above the threshold the growth rate is given by

$$\gamma^2 = \tfrac{1}{2}\{\delta^2 + k_s^2 c_s^2 - [(\delta^2 - k_s^2 c_s^2)^2 - 4K\delta]^{1/2}\} \tag{126}$$

$$\delta = \omega_0 - \omega_1.$$

For $\delta \gg \omega_s$ we obtain

$$\gamma = \frac{1}{\sqrt{2}} m_{pi} v_0/c, \qquad v_0 = \frac{e|E_{t_0}|}{m_e \omega_0}. \tag{127}$$

The minimum threshold of the filamentation instability is given by

$$v_0^2/c_s^2 = 8\omega_1 \Gamma_T/\omega_{pi}^2 \tag{128}$$

and the growth rate by

$$\gamma = \omega_{pi} v_0/\sqrt{2}c \tag{129}$$

where Γ_T is the damping rate of transverse waves.

NONLINEAR TREATMENT OF FILAMENTATION

In a manner analogous to the treatment of the nonlinear evolution of the three wave decay, the filamentation can be analysed nonlinearly. From Fig. 16, where k_0 and k_s are seen to be nearly perpendicular, we deduce from the vector relation

$$\mathbf{k}_1 = \mathbf{k}_0 + \mathbf{k}_s, \qquad \mathbf{k}_2 = \mathbf{k}_0 - \mathbf{k}_s \tag{130}$$

that

$$k_{1,2}^2 \approx k_0^2 + k_s^2. \tag{131}$$

We also have the dispersion relations

$$c^2 k^2 = \omega^2 - \omega_p^2. \tag{132}$$

The mismatch δ may be calculated by means of (131) and (132); we have

$$-c^2 k_s^2 = c^2(k_0^2 - k_1^2) = \omega_0^2 - \omega_1^2 \approx (\omega_0 - \omega_1)2\omega_0 = 2\omega_0\delta. \tag{133}$$

It is instructive to compute the wave energy in terms of the field amplitudes $\mathbf{E}, \mathbf{H}$,

$$\varepsilon = \{\tfrac{1}{4}\mu_0 \mathbf{H}.\mathbf{H}^* + \tfrac{1}{4}\varepsilon_0 \mathbf{E}.\mathbf{E}^* + \tfrac{1}{4}n_0 m_e \mathbf{v}.\mathbf{v}^*\}_{\omega, k}. \tag{134}$$

For the three transverse waves. Eliminating E by Maxwell's equation and expressing the fluid velocities by $v = eE_z/m\omega$ we find

$$\varepsilon = \left\{ \left(1 + \frac{c^2 k^2}{\omega^2} + \frac{\omega_{pe}^2}{\omega^2} \right) \tfrac{1}{4}\varepsilon_0 E_z^2 \right\}_{\omega,\,k} \tag{135}$$

and in view of the dispersion relation for the pump wave this becomes

$$\varepsilon_{\text{pump}} = \tfrac{1}{2}\varepsilon_0 |E_{T_0}|^2. \tag{136}$$

The Stokes and anti-Stokes wave-energy is evaluated at $k_{1,\,2}$ and we obtain

$$\varepsilon_{1,\,2} = \left(1 - \frac{\delta}{\omega_0} \right) \tfrac{1}{2}\varepsilon_0 |E_{T_{1,\,2}}|^2. \tag{137}$$

It is convenient to introduce variables

$$\alpha_0 = \left(\frac{\varepsilon_0}{2} \right)^{1/2} E_{T_0};\; \alpha_1 = \left(\frac{\varepsilon_0}{2} \right)^{1/2} E_{T_1} e^{i\delta t}, a_2 = \left(\frac{\varepsilon_0}{2} \right)^{1/2} E_{T_2}^* e^{-i\delta t} \tag{138}$$

and in terms of these the usual three equations become

$$\frac{\partial \alpha_0}{\partial t} = \frac{2i\Gamma}{\varepsilon_0 \omega_0} (|\alpha_1|^2 \alpha_0 + (\alpha_2)^2 \alpha_0 + 2\alpha_0^* \alpha_1 \alpha_2) \tag{139}$$

$$\left(\frac{\partial}{\partial t} - i\delta \right) \alpha_1 = \frac{2i\Gamma}{\varepsilon_0 \omega_0} (|\alpha_0|^2 \alpha_1 + |\alpha_0|^2 \alpha_2) \tag{140}$$

$$\left(\frac{\partial}{\partial t} - i\delta \right) \alpha_2 = \frac{2i\Gamma}{\varepsilon_0 \omega_0} (|\alpha_0|^2 \alpha_2 + |\alpha_0^*|^2 \alpha_1). \tag{141}$$

It is not difficult to show by means of these equations that

$$\frac{\partial}{\partial t} \left\{ |\alpha_0|^2 + \left(1 - \frac{\delta}{\omega_0} \right) |\alpha_1|^2 + \left(1 - \frac{\delta}{\omega_0} \right) |\alpha_2|^2 \right\} = 0 \tag{142}$$

which clearly express energy conservation, as the three terms are the time changes of the three wave energies. Since all the waves have frequency ω_0 dividing by it we find the action conservation equation which is equivalent to the Manley–Rowe equations for the three-wave decay.

As before, an exact solution of (139–141) is possible in terms of elliptic functions. The details may be found in [12]. We illustrate the time-dependent solution in Figs 17, 18 where the three waves and the associated density are plotted for certain values of the normalized mismatch parameter

$$\Delta = \frac{4_c^2 k_s^2}{\omega_{pe}^2} \frac{v_{T_0}^2}{v_0^2} \tag{143}$$

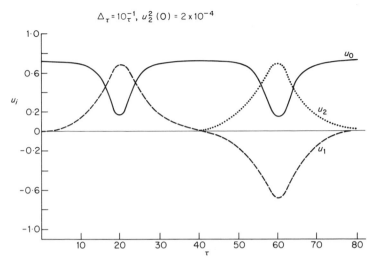

Fig. 17. Solutions of the nonlinear filamentation problem. Normalized amplitude of the incident wave (u_0; full curve) the Stokes wave (u_1; broken curve) and anti-Stokes wave (u_2; dotted where it is different from former.

and a certain (low) initial amplitude of the anti-Stokes wave. The normalized variable τ is $\tau = \frac{1}{4}(v_0^2/v_{T_2})(\omega_{pe}^2/\omega_1^2)\omega_0 t$. It can be seen that there is a maximum density depletion in the first half cycle when the Stokes and anti-Stokes waves have the same sign and a maximum density accumulation in the second half cycle when the two amplitudes are opposite in sign. Maxima and minima of density and excited waves occur at the same time.

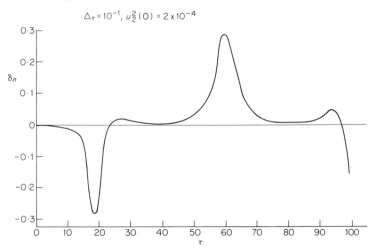

Fig. 18. The density perturbation associated with filamentation (as a fraction of the equilibrium density) plotted as a function of normalized time τ for $\Delta = 10^{-1}$ and $u(0)^2 = 2 . 10^{-4}$.

STATIONARY SOLUTIONS

Again we assume solutions to exist which depend on space and time in the combination $\xi = x - ut$, wave patterns which are transported with velocity u.
    The algebra is rather involved and finally leads to an equation

$$\frac{dU}{d\xi} = -K[(U - U_1)(U - U_2)(U_3 - U)]^{1/2} \tag{144}$$

for the variable $U = a_1^2(\xi)$ where we have put

$$E_{T_j}(\xi, t) = a_j(\xi, t)\exp\left[i(\phi_j - 2\delta_j t)\right] \tag{145}$$

where $\delta_1 = \delta_2 = 0$ and $j = 1, 2, 3$.

    The quantities $U_{1,2,3}$ depend on the plasma parameters. Equation (144) is analogous to that for the motion of a particle in a potential $(U - U_1)$ $\times (U - U_2)(U_3 - U)$. The solution is found in terms of elliptic functions

$$a_1^2(\xi) = U_2 + (U_3 - U_2)c_n^2(\beta\xi, k) \tag{146}$$

and the other two amplitudes involve s_n and d_n. Here

$$\beta = \frac{K}{2}(U_3 - U_1)^{1/2}; \qquad k^2 = \frac{U_3 - U_2}{U_3 - U_1}.$$

For $U_1 = U_2$, $k = 1$ and we have solitary wave solutions proportional to $\mathrm{sech}(\beta\xi)$ and $\tanh(\beta\xi)$. The curves are slowly varying amplitudes of high frequency waves which undergo many oscillations within them. They are called envelope solitons. (See Fig. 14 which shows this shape of the envelopes.)

REFERENCES

1. Einstein, A. (1917). *Physik. Z.* **18**, 121.
2. Peierls, R. E. (1955). "Quantum Theory of Solids", p. 44. Oxford University Press, Oxford.
3. Tystovich, V. N. (1970). "Nonlinear Effects in Plasma". Plenum Press, London.
   Tystovich, V. N. (1972). "An Introduction to the Theory of Plasma Turbulence". Pergamon Press, Oxford.
   Tystovich, V. N. (1976). *In* "Proc. 12th Int. Conf. on Phenomena in Ionized Gases, Eindhoven". North Holland Publishing Co., Amsterdam.
4. Motz. H. (1955). *In* "Symposium on Electromagnetic Theory", Trans. I.R.E., AP 4, 374.
5. Weiland, J. and Wilhelmsson, H. (1977). "Coherent Nonlinear Interaction of Waves in Plasma". Pergamon Press, Oxford.
6. Salpeter, E. E. (1961). *Phys. Rev.* **122**, 1663.

7. Evans, D. E. and Katzenstein, J. (1969). *Rep. Prog. Phys.* **32**, 207–271.
8. Buneman, O. (1959). *Phys. Rev.* **115**, 503–517.
9. Nishikawa, K. (1968). *J. Phys. Soc. Japan*, **24**, 916, 1152.
10. Franklin, R. N. (1977). *Rep. Prog. Phys.* **40**, 1369–1413.
11. Hamberger, S. M. and Clark, W. H. M. (1975). Seventh Europ. Conf. on Controlled Fusion and Plasma Physics, Lausanne, CRPP, Ecole Polytechnique.
12. Bingham, R. (1977). Ph.D. Thesis, Oxford University.
    Bingham, R. and Lashmore Davies, C. N. (1979). *Plasma Phys*, **21**, 433.
13. Bingham, R. and Lashmore Davies, C. N. (1976). *Nucl. Fus.* **16**, 1, 67–72.
14. Krylov, N. M. and Bogoliubov, N. N. (translated by S. Lefschetz) (1943). "Introduction to Nonlinear Mechanics". Princeton University Press, Princeton, N.J.
15. Chu, L. J. (1951). Proc. I.R.E. Conf. Dublin, unpublished.
    Sturrock, P. A. (1960). *J. Appl. Phys.* **31**, 2052.
16. Perkins, F. W. and Flick, J. (1971). *Phys. Fluids*, **14**, 2012.
17. Rosenbluth, M. N., Liu, C. S. and White, R. B. (1974). *Phys. Fluids*, **17**, 1211–1219.
18. Jahnke, E. and Emde, F. (1945). "Tables of Functions with Formulae and Curves". Dover Publications, New York.
19. Manley, J. M. and Rowe, H. E. (1959). *Proc. IRE*, **47**, 2115.
20. Volkov, T. F. (1960). *In* "Plasma Physics and the Problems of Controlled Thermonuclear Reactions" (M. A. Leontovich, ed.), Vol IV, p. 114. Pergamon Press, London.
21. Silin, V. P. (1965). *Sov. Phys. JETP*, **21**, 1127.
22. Goldman, M. V. and DuBois, D. F. (1975). *Phys. Fuids*, **8**, 1404.
23. Motz, H. (1975). *Ann. New York Acad. Sci.*, **251**, 74.

6

QUASI-POTENTIAL AND PONDEROMOTIVE FORCE

We are concerned with laser light of large amplitude so that radiation pressure effects are not negligible. In many situations they become so large, that the self-consistent field pressure can balance the particle pressure. Plasma layers are trapped or confined between field layers; the field in turn cannot penetrate the plasma region if the plasma frequency exceeds the r-f field frequency. (We use the term r-f, (radio frequency), for all electromagnetic fields because our considerations apply equally well to microwaves etc.)

We shall first study single particle motions in r-f fields and we shall find that the equation of motion averaged over the r-f time variation possesses an integral which is a constant of the slow motion. This fact will allow us to study self-consistent problems, i.e. situations where the plasma distribution is determined by the field distribution, while the field, in turn, is determined by the plasma distribution. We shall show that the particle distribution is a function of the constants of motion in which the field figures.

The field will be inhomogeneous, with length scale of inhomogeneity $L_E = |(\nabla E)/E|^{-1}$. *In vacuo,* for a field with spatial variation $\exp ikx$, the length scale is $L_V = k^{-1} = c/\omega$. In first order, the velocity v_E of a particle acquired in a field $\mathbf{E}$ is given by $v_E = eE/m\omega$ called the quiver velocity and its displacement $\rho_E = e\mathbf{E}/m\omega^2$. The ratio of the energy $mv_E^2/2$ and the thermal energy $kT/2$ in one degree of freedom is $e^2E^2/m\omega^2kT$ and this is a relevant measure of the r-f energy which determines whether the confinement effects become important.

The equation of motion of a single particle will be formulated in a system of units in which $e = m = 1$ and the fields $\mathbf{E}$ and $\mathbf{B}$ both have the dimensions of acceleration. The algebra then becomes more transparent. The results can be converted into the mks system by replacing $\mathbf{E}$ by $e\mathbf{E}/m$ and putting $c = 1$.

In this system of units the equation of motion is

$$\ddot{\mathbf{r}} = \mathbf{E}(\mathbf{r}, t) + \mathbf{v}/c \times \mathbf{B}(\mathbf{r}, t). \tag{1}$$

We write $\mathbf{r}$ as a sum of a smoothly varying, or guiding centre motion $\mathbf{R}$ and a rapidly oscillating motion $\rho(t)$.

Substituting $\mathbf{r} = \mathbf{R} + \mathbf{\rho}$ into (1) expanding for small ρ and separating into smooth and oscillating parts we obtain

$$\ddot{\mathbf{R}} = \langle(\mathbf{\rho} \cdot \mathbf{V})\mathbf{E}(\mathbf{R}, t)\rangle + \left\langle \frac{1}{c}(\mathbf{\rho} \times \mathbf{B}) \right\rangle \tag{2}$$

$$\ddot{\mathbf{\rho}} = \mathbf{E}(\mathbf{R}, t) + [\mathbf{\rho} \cdot \mathbf{V}\mathbf{E} - \langle\mathbf{\rho} \cdot \mathbf{V}\mathbf{E}\rangle] + \frac{1}{c}[(\dot{\mathbf{\rho}} \times \mathbf{B}) - \langle(\dot{\mathbf{\rho}} \times \mathbf{B})\rangle] + \frac{1}{c}(\dot{\mathbf{R}} \times \mathbf{B}) \tag{3}$$

where $\langle A \rangle$ is a quantity obtained from A by averaging its explicit time dependence over a time $2\pi/\omega$. In the second term of (3) the rapid motion term was obtained by subtracting the average. This second term is of order

$$\frac{v_E}{\omega} \frac{\nabla E}{E} = \frac{v_E}{c}(L_V/L_E)$$

relative to the first one, E. Our theory will be valid when the scale-length ratio $L_V/L_E < 1$ so that, if we neglect terms in (3) other than E we make an error of the order v/c. Furthermore, the error made in ignoring the implicit time-dependence of E (through $R(t)$) when integrating (3) is also of higher order in v/c.

We assume that the field is represented by

$$\mathbf{E} = \mathbf{E}e^{i\omega t} + \mathbf{E}^*e^{-i\omega t} \tag{4}$$

and we have

$$\dot{\mathbf{\rho}} = \frac{\mathbf{E}e^{i\omega t} - \mathbf{E}^*e^{-i\omega t}}{i\omega}, \mathbf{\rho} = -\frac{\mathbf{E}(E, t)}{\omega^2}. \tag{5}$$

On inserting these into (2) and using Maxwell's equation

$$\frac{1}{c}\mathbf{B}(R, t) = -\mathbf{V} \times \int^t \mathbf{E}(R, t')\,\mathrm{d}t' \tag{6}$$

we obtain

$$\ddot{\mathbf{R}} = \frac{-1}{\omega^2}[\mathbf{E} \cdot \mathbf{V}\mathbf{E}^* + \mathbf{E}^* \cdot \mathbf{V}\mathbf{E} + \mathbf{E} \times \mathbf{V} \times \mathbf{E}^* + \mathbf{E}^* \times \mathbf{V} \times \mathbf{E}]$$

$$= -\mathbf{V}\mathbf{E} \cdot \mathbf{E}^*/\omega^2 = -\mathbf{V}\psi. \tag{7}$$

Thus the smooth guiding centre motion is derivable from a scalar potential
ψ: in order to avoid implying any judgement on priorities in a field where
Soviet, British and American candidates all possess claims (Gaponov and
Miller, Boot and Shersby Harvie, Weibel), we shall refer to ψ as the "quasi-
potential" for the guiding centre. The first integral of the motion, re-expressed
in dimensional form is

$$\xi = \tfrac{1}{2}m\dot{R}^2 + \frac{e^2|E|^2}{m\omega^2} = \text{constant}. \tag{8}$$

The apparent discrepancy in numerical factor between this and various other
expressions for the quasi-potential to be found in the literature [1] is due to
different definitions of the relationship of the complex quantities E and E^*
to the real magnitude of the electric field, which may be represented by its
r.m.s. or peak value. In terms of the peak value E_p, $\psi = e^2|E|_p^2/4m\omega^2$; in
terms of the r.m.s. value $E_{r.m.s.}$, $\psi = e^2|E|^2 r.m.s./2m\omega^2$. In this book we shall
use the r.m.s. value throughout, since with this choice, we can also write
$\psi = e^2|E|^2/2m\omega^2$ where $E(R, t)$ is the instantaneous value of the electric field
at the guiding centre. An additional factor of one half arises when one comes
to impose the condition of quasi-neutrality in a plasma, as we shall see below.
Expressing (7) in dimensional form

$$m\ddot{R} = -\nabla\psi; \qquad \psi = e^2|E|^2/m\omega^2 \tag{9}$$

we see that $\nabla\psi$ is the average force acting on a time scale large compared
to $2\pi/\omega$. It is called the ponderomotive force in the literature on the subject.
This force acts on particles in a plasma penetrated by an e.m. field. The field
distribution in the plasma however must be calculated from Maxwell's
equations, i.e. the wave equation

$$\nabla \times \nabla \times E = -\frac{1}{c^2}\frac{\partial}{\partial t}\left(\frac{\partial E}{\partial t} + j\right) \tag{10}$$

and

$$-\nabla^2\phi - \nabla \cdot E = \rho/\varepsilon_0 \tag{11}$$

where the charge density and current density are given by

$$\rho = e\int(f_+ - f_-)\,d^3v \tag{12}$$

$$j = e\int v(f_+ - f_-)\,d^3v. \tag{13}$$

In a quasi-neutral plasma, charge separation can exist, giving rise to an
electrostatic potential ϕ which is determined by Poisson's equation. The
integral of motion must be modified by inclusion of this potential so that

$$\tfrac{1}{2}m_\pm\dot{R}^2 + \psi_\pm \pm e\phi = \xi_\pm = \text{constant} \tag{14}$$

an estimate of this potential ϕ will be found below. For a collisionless plasma the distribution functions $f_{\pm}$ must satisfy the Vlasov equation. They can be chosen as arbitrary functions of the constants of motion $\xi_{\pm}$. Indeed substituting $f_{\pm}(\xi_{\pm})$ in the Vlasov equation we find, putting $v = \dot{R}$

$$\frac{\partial f_{\pm}}{\partial \xi_{\pm}} \mathbf{v} \cdot \mathbf{\nabla} \xi_{\pm} - \frac{1}{m} \mathbf{\nabla}(\psi \pm e\phi) \cdot \frac{\partial \xi_{\pm}}{\partial \mathbf{v}} \frac{\partial f_{\pm}}{\partial \xi_{\pm}} = 0.$$

We shall normalize the distribution functions to be equal to unity at points where $\psi = \phi = 0$. From (10) to (13) we then find (with $\partial \mathbf{v}/\partial t = e\mathbf{E}/m$ neglecting the slow time-scale term).

$$\mathbf{\nabla} \times \mathbf{\nabla} \times \mathbf{E} = \frac{\omega^2 \mathbf{E}}{c^2}\left(1 - \frac{\omega_{pi}^2}{\omega^2}\int f_+ \, d^3\mathbf{v} - \frac{\omega_{pe}^2}{\omega^2}\int f_- \, d^3v\right) = \frac{\omega^2}{c^2}\varepsilon \mathbf{E} \qquad (15)$$

$$\mathbf{\nabla}\cdot\mathbf{E} - \nabla^2\phi = e/\varepsilon_0\left[n_{0+}\int f_+ \, d^3\mathbf{v} - n_{0-}\int f_- \, d^3\mathbf{v}\right] \qquad (16)$$

where ω_{pi}, ω_{pe} are the plasma frequencies $(e^2 n_{0-}/m\varepsilon_0)^{1/2}$, $(e^2 n_{0+}/m\varepsilon_0)^{1/2}$ at $\psi = \phi = 0$. It is clear that in our approximation ρ does not have a high frequency component and that $\mathbf{\nabla}\cdot\mathbf{E} = 0$. In dimensionless form (16) is written as

$$-\nabla^2 e\phi/kT = \frac{\omega_{pi}^2}{kT/m_i}\int f_+ \, d^3\mathbf{v} + \frac{\omega_{pe}^2}{kT/m_e}\int f_- \, d^3\mathbf{v}. \qquad (17)$$

We assume that the variation of ϕ has the same length scale c/ω as that of the quasi-potential, and we see that the ratio of the increment $\Delta e\phi$ to kT is small compared to unity provided that $(\omega^2/\omega_{pe}^2) v_T^2/c^2 \ll 1$. Thus we can take $\nabla^2\phi = 0$ unless the plasma density is very low. In other words, quasi-neutrality is consistent with quasi-potential theory. This means that for $T_e - T_i$

$$\int f_+(\tfrac{1}{2}mv_+^2 + \psi_+ + e\phi) \, d^3\mathbf{v} \approx \int f_-(\tfrac{1}{2}mv_-^2 + \psi_- - e\phi) \, d^3\mathbf{v}$$

and if f_+ and f_- are the same functions of their different arguments (e.g. Maxwellians) we have

$$2e\phi = \psi_- - \psi_+. \qquad (18)$$

Neglecting terms of the order of m_e/m_i compared with unity we have

$$\mathbf{\nabla} \times \mathbf{\nabla} \times \mathbf{E} = \frac{\omega^2}{c^2}\varepsilon \mathbf{E} = \frac{\omega^2}{c^2}\left[1 - \frac{\omega_{pe}^2}{\omega^2}\int f\left(\tfrac{1}{2}m_e v_e^2 + \frac{e^2 E^2}{4m_e\omega^2}\right) d^3\mathbf{v}\right]\mathbf{E} \qquad (19)$$

where $\mathbf{E}$ still has the time dependence $\exp i\omega t$.

ONE-DIMENSIONAL EQUILIBRIA

For plane waves, equation (19) takes the form

$$\frac{d^2\mathbf{E}}{dz^2} + \frac{\omega^2}{c^2}\varepsilon\mathbf{E} = 0 \qquad \text{where} \quad \mathbf{E} = (E_x, E_y, 0) \tag{20}$$

in general, this is a pair of coupled equations for E_x and E_y. However if we take the case of circularly polarized waves so that $\mathbf{E}(z, t) = \mathscr{E}(z)\mathbf{e}(\omega t)$ where $\mathbf{e}$ is a periodic function of ωt such that $\langle e^2 \rangle = 1$, (20) reduces to

$$\frac{d^2\mathscr{E}}{dz^2} + \frac{\omega^2}{c^2}\varepsilon\mathscr{E} = 0$$

$$\varepsilon = 1 - \frac{\omega_p^2}{\omega^2}\int f\left(\tfrac{1}{2}m_e v_e^2 + \frac{e^2\mathscr{E}^2}{4m_e\omega^2}\right)d^3\mathbf{v}. \tag{21}$$

If one further chooses f to be Maxwellian this can be written as

$$\frac{d^2E}{dz^2} + \frac{\omega^2}{c^2}E = \frac{\omega_p^2}{c^2}E\exp\left(-\frac{-e^2\mathscr{E}^2}{4m\omega^2 kT}\right) \tag{22}$$

Note that, from Maxwell's equations $dE/dz = \omega B(z)$ and that $c^2 = 1/\mu_0\varepsilon_0$. It is instructive to write

$$\omega_p^2 E\exp\left(\frac{-e^2\mathscr{E}^2}{4m\omega^2 kT}\right) = -2n_0\omega^2(kT/\varepsilon_0)\frac{d}{dE}\exp\left(\frac{-e^2\mathscr{E}^2}{4m\omega^2 kT}\right)$$

and to multiply (22) by dE/dz and to integrate with respect to z. The result is

$$\tfrac{1}{2}(\varepsilon_0\mathscr{E}^2 + \mu_0\mathscr{H}^2) + 2n_0 kT\exp\left(\frac{-e^2\mathscr{E}^2}{4m\omega^2 kT}\right) = \text{constant} \tag{23}$$

or

$$\tfrac{1}{2}(\varepsilon_0\mathscr{E}^2 + \mu_0\mathscr{H}^2) + p = \text{constant} \tag{23a}$$

and it shows that the plasma pressure $p = 2n_0 kT\exp(-e^2\mathscr{E}^2/4m\omega^2 kT)$ balances the radiation pressure, where $2n_0\exp(-e^2\mathscr{E}^2/4m\omega^2 kT)$ is the plasma density $n_e + n_i$ at a position where the field strength is E.

We can also write (23) in the form

$$\left(\frac{d\mathscr{E}}{dz}\right)^2 + \frac{\omega^2}{c^2}\left[\mathscr{E}^2 + \frac{4\omega_p^2 kTm}{e^2}\exp\left(\frac{-e^2\mathscr{E}^2}{4m\omega^2 kT}\right)\right] = \text{constant} \tag{24}$$

To solve this for E as a function of z we need another integration. We can however infer the essential features of the solution by considering a mechanical analogy (see [1, 2]). For simplicity we shall measure z on the scale c/ω

and $\mathscr{E}$ on the scale $(4m\omega^2 kT/e^2)^{1/2}$. We can rewrite (24) as

$$\left(\frac{d\mathscr{E}}{dz}\right)^2 + \psi(\mathscr{E}) = \text{constant} = C^2 \tag{25}$$

with

$$\psi(\mathscr{E}) = \mathscr{E}^2 + \frac{\omega_p^2}{\omega^2} e^{-E^2}$$

This is analogous to the first integral of motion of a mechanical particle of twice unit mass in a potential well $\psi(E)$ when the correspondences

$$z \to t \,\text{(time)}$$

$$E \to x \,\text{(particle coordinate)}$$

$$dE/dz \to \text{velocity}$$

are noted. A full discussion has been given in [1] but we shall highlight some conclusions. Figure 1 shows the shape of the potential surface for (i) $\omega_p^2/\omega^2 < 1$ and for (i) $\omega_p^2/\omega^2 > 1$.

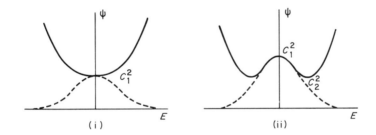

Fig. 1. Mechanical analogue of confinement problem. Potential surface (full curve) plasma density (broken curve) plotted against field strength. A ball rolling on this surface represents the system if E is interpreted as its coordinate and the time when it reaches a given position is the spatial coordinate of the field configuration. (i) shapes for $\omega_p^2/\omega^2 > 1$; (ii) shapes for $\omega_p^2/\omega^2 > |$.

The figure shows, in broken lines the function $\omega_p^2/\omega^2 \, e^{-E^2}$ (which is a measure of the plasma density as a function of field) rising to a maximum $C_1^2 = \omega_p^2/\omega^2$. The full curve represents $\psi(E)$. The total particle energy is C^2.

First we consider case (i). Keeping the mechanical analogy in mind, the particle remains shakily at rest when $C = C_1$. If C is not much larger than C_1 the shape of the potential is parabolic and the particle carries out harmonic oscillations as a function of time: the field strength varies harmonically with distance. The plasma density is a maximum where E vanishes, i.e. at the position where the r-f field does not penetrate.

When C is made still higher, i.e. when the particle is released from a position on the ψ curve situated much above C_1 its movement becomes non-sinusoidal. We still have layers of r-f field separated by plasma layers. The plasma is confined by the field, it is overdense near the maximum density so that the r-f field is excluded.

Case (ii) also has a special solution with $C \approx C_1$. In this case the solution is unstable, a small displacement makes the particle slide down the hill, first very slowly, then faster and then after reaching the bottom, climb up the other side; there is a turning point, the particle comes back, climbs up, but near $E = 0$ moves very slowly, and takes an infinite time to come to rest. The curve of E vs z is shown in Fig. 2; it is a solitary standing wave. If $C_1 > C$

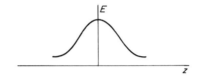

Fig. 2. Solitary wave solution arising when $C \gtrsim C_1$.

$> C_2$ we have non-sinusoidal oscillations, but the field strength remains positive, or negative (Fig. 3). For $C > C_1$ again we have a non-sinusoidal oscillation. We shall not continue the discussion detail, it is clear that the exact curves can be obtained by computation.

The curve of Fig. 2 represents a so-called stationary solitary wave, a hump-like solution of a type which will be further discussed below. It represents an r-f pocket trapped in plasma, and held together by the plasma pressure.

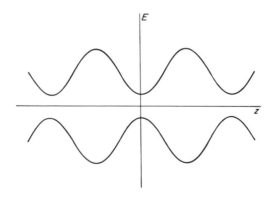

Fig. 3. Non-sinusoidal solution of radiation confinement equation arising for $C_1 > C > C_2$.

A General Expression for the Nonlinear Force

A general expression for the nonlinear force per unit volume f_{NL} can be given in terms of the refractive index $n_r = k/k_0$ of a transverse wave propagating in a plasma with space–time dependence $\exp(i\omega t - n_r k_0 z)$ where k_0 is the vacuum propagation constant. It may be written in the form

$$f_{NL} = \frac{\varepsilon_0}{2} (n_r^2 - 1) \nabla |E|^2 \qquad (26)$$

for linearly polarized waves (cf. [13]).

This can be shown starting with the relation (23) written (remembering that *cursive letters* denoted amplitudes)

$$-\frac{1}{2} \frac{\partial}{\partial z} [\varepsilon_0 |E|^2 + \mu_0 |H|^2] = \frac{\partial q}{\partial z} = f_{NL}$$

It is assumed that the spatial scale of inhomogeneity $(1/n_r) \partial n_r / \partial z$ of n_r is low enough to allow a representation of the field in the form

$$E = E_0 n_r^{-1/2} \exp\left(-ik \int n_r \, dz\right)$$

(see Budden [7], Chapter 7 on the WKB method.)

From Maxwell's equations, it is easy to show that for linearly polarized wave propogation in the z-direction

$$|H|^2 = n_r^2 Y_0^2 |E|^2 \qquad Y_0 = \left(\frac{\varepsilon_0}{\mu_0}\right)^{1/2}.$$

We can thus write

$$f_{NL} = -\frac{\varepsilon_0}{2} \frac{\partial}{\partial z} [|E|^2 (1 + n_r^2)] =$$

$$-\frac{\varepsilon_0}{2} \left[(1 + n_r^2) \frac{\partial |E|^2}{\partial z} + |E|^2 \, 2n_r \frac{\partial n_r}{\partial z} \right]$$

$$-\frac{\varepsilon_0}{2} \left[(1 - n_r^2) \frac{\partial |E|^2}{\partial z} \right] - \frac{\varepsilon_0}{2} \left[2n_r^2 \frac{\partial |E|^2}{\partial z} + 2n_r \frac{|E|^2 \, \partial n_r}{\partial z} \right]$$

But the second term equals $\partial/\partial z (2|E|^2 n_r)$ which vanishes since $|E|^2 n_r = E_0^2 = $ constant.

For transverse waves $n_r^2 = 1 - \omega_p^2/\omega^2$ and we can write

$$f_{NL} = -\frac{\omega_p^2 \varepsilon_0}{\omega^2} \nabla |E|^2. \qquad (26a)$$

Here E is the field in the plasma. In terms of the vacuum field (26) is written in the form

$$f_{NL} = -\frac{\varepsilon_0 E_0^2}{2} \frac{\omega_p^2}{\omega^2} \nabla \frac{1}{n_r} \tag{26b}$$

and it it seen that there is a singularity at $\omega = \omega_p$.

EXPLANATION OF THE PHYSICAL MECHANISM OF THE OSCILLATING TWO-STREAM INSTABILITY

As an application of the concept of a ponderomotive force we return to the subject of the OTS discussed before. An electromagnetic wave with dispersion equation

$$\omega_0^2 = \omega_p^2 + c^2 k_0^2 \tag{27}$$

decays into a Langmuir wave and an acoustic wave with a frequency $\omega_s \ll \omega_p$. The behaviour is different for $\omega_0 < \omega_p$ and $\omega_0 > \omega_p$. This is indicated by the sign of δ in Chapter 5, formula (79). In the former case an instability arises, the OTS, which has zero real frequency but grows in time. Since $\omega_s \ll \omega_p$, $\omega_0 \approx \omega_p$ and $k_0 \approx 0$. We may regard E_0 as spatially uniform and we shall now explain how the instability arises with reference to Fig. 4. We presume

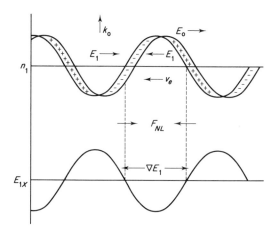

Fig. 4. Illustration of the physical mechanism of the oscillating stream instability.

that there is a density perturbation of the form $n_1 \sin k_1 x$ arising from thermal noise. If $\omega_0 < \omega_p$, electrons will move in a direction opposite to E_0, the pump field, and the ions are too inert to follow the r-f field. Hence charge separation will occur as indicated in the figure. This will cause an electric field E_1 oscillating at frequency ω_0 to appear. The total field is $E = E_0 + E_1$ and the quasipotential due to this field is given by

$$\psi = \frac{e^2 \langle E^2 \rangle}{2m\omega^2}. \tag{28}$$

The force on a particle is $F = -\nabla\psi$, or, if the density is n, the ponderomotive force per unit volume is $-ne^2\nabla\langle\varepsilon_0 E^2\rangle/2m\varepsilon_0\omega^2$. However, $E_0 \gg E_1$ and only the term

$$-\frac{e^2 n}{2m\omega^2}\nabla\langle E_0 E_1\rangle, \qquad n = n_0 + n_1 \tag{29}$$

is important. The time average $\langle E_0 E_1\rangle$ does not vanish because E_1 changes size with E_0. The force vanishes at the maxima and minima of the density distribution but is largest where the density gradient is maximal. Thus, as indicated in Fig. 4 the ponderomotive force will tend to push electrons from low density regions into high density regions and the resulting d.c. field will drag the ions along too. Thus the density perturbation grows as long as F is large enough to overcome the particle pressure. This density ripple does not propagate and so $\text{Re}(\omega_1) = 0$.

When $\omega_0 > \omega_p$ this mechanism does not work because an oscillator driven faster than the resonant frequency moves opposite to the direction of the applied force. The direction v_e and E_1 are reversed so that the ponderomotive force now moves electrons from dense into less dense regions. The perturbation dies out. The phases leading to instability can be restored if the density perturbation moves with a finite velocity. Let the disturbance now move with the acoustic velocity v_s. Then in the frame moving with the density perturbation electrons are subjected to a pump field at the Doppler-shifted frequency $\omega' = \omega_0 - k_1 v_s$ and if this frequency equals the plasma frequency then the interaction is resonant and therefore large. This gives $\omega' = \omega_p = \omega_0 - k_1 v_s$ but the ion wave moves with phase velocity $\omega_1/k_1 = v_s$, $\omega_1 = k_1 v_s$ and we get $\omega_p = \omega_0 - \omega_1$ which is just the frequency matching condition relating the Langmuir frequency, the transverse oscillation and the acoustic frequency we already know.

In a paper by Drake et al. [3] parametric instability theory is based on the effect of the ponderomotive force.

Solitons

We have already encountered solitary wave solutions in connection with the nonlinear evolution of modulation and filamentation and the r–f confinement equation (24). There is a class of solitary waves which is of great interest and has recently attracted much attention: solitons.

The first observation of a soliton was recorded by Scott Russell who saw one while riding along a waterway (see bibliography in [4]). In that case, the water wave observed obeys a nonlinear wave equation. Solutions of a wave equation which depends on x and t only through $\xi = x - ut$, where u is a constant, form a class of travelling waves from which one can pick out a subclass, localized solutions, whose transition from one constant amplitude state at $\xi \to -\infty$ to another as $\xi \to +\infty$ is essentially localized in x. Figure 2 shows an example of such a wave with zero speed. Figure 5 is another, the derivative of which looks like the first.

Fig. 5. Solitary wave resembling a shock profile.

Perring and Skyrme [5] noticed that two solitary waves which were solutions of a certain equation emerged from a collision with the same shapes and velocities they had before collision. Shortly afterwards Zabusky and Kruskal [7] independently found the same remarkable property by means of computer studies of the equation

$$\psi_t + \alpha\psi_x + \beta\psi_{xxx} = 0 \tag{30}$$

the so-called Korteweg de Vries (KdV) equation. They coined the term soliton for waves with this property. Equation (30) describes the asymptotic behaviour, for long times, of the equations of ion-acoustic waves, water waves and other physical situations. Given any solution $\phi(x, t)$ composed of solitary waves ϕ_{sj} at large negative times

$$\phi(x, t) \approx \sum_{j=1}^{N} \phi_{sj}(\xi_j), \qquad t \to -\infty, \tag{31}$$

$$\xi_j = x - v_j t, \qquad v_j \text{ constant,}$$

such solitary waves will be called solitons if they emerge from an interaction with no more change than a phase shift, i.e.

$$\phi(x, t) \approx \sum_{j=1}^{N} \phi_{sj}(\bar{\xi}_j) \quad \text{as} \quad t \to +\infty \tag{32}$$

$$\bar{\xi}_j = x - v_j t + \delta_j, \qquad \delta_j \text{ constant.}$$

This is illustrated by Fig. 6 taken and redrawn from a paper by Ikezi *et al.* who actually observed this behaviour in the case of ion acoustic waves. The solitons are seen to approach, merge, and re-separate. The pulse solutions of a *linear* wave equation have this property trivially, but solutions of nonlinear equations can have it only if they are also dispersive.

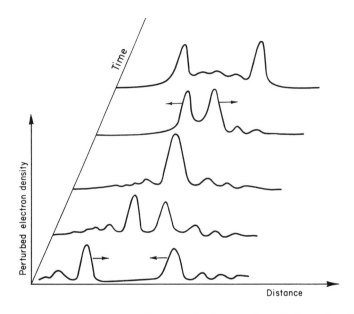

Fig. 6. Non-destructive collision of ion acoustic plasma pulses. (Redrawn from [8].)

Dispersion

We have had many examples of dispersion equations arising out of the theory of plasma waves. Such a relation

$$\omega = W(k) \tag{33}$$

between frequency and wave number characterized many problems. The form

$$\psi = \phi \exp[i(\omega t - \mathbf{k} \cdot \mathbf{x})] \tag{34}$$

inserted in the KdV equation (30) leads to the dispersion equation

$$\omega = \alpha k - \beta k^3. \tag{35}$$

We shall also consider the so-called nonlinear Schrödinger equation

$$\psi_{xx} + i\psi_t + \beta|\psi|^2\psi = 0, \qquad \beta > 0. \tag{36}$$

Here the dispersion equation   (37)

$$\omega = -k^2 + \beta A$$

depends on the amplitude $A = |\psi|^2$. This is also an example of an equation with mixed derivatives. (The factor i leads to a real dispersion equation although the derivatives are neither all even or all odd).

We note that the phase velocities

$$\omega/k = \alpha - \beta k^2, \tag{38}$$

$$\omega/k = A\beta/k - k, \tag{39}$$

change with wave number. This tends to flatten out, or disperse, steep wave fronts. We shall see in detail how most nonlinear equations have solutions which steepen until shocks from. A simple example, equation

$$\psi_t + c(\psi)\,\psi_x = 0 \tag{40}$$

is discussed in the chapter on fluid dynamics.

Our solitons owe their stability, their constant shape, to the balance between the dispersive tendency and the steepening tendency of nonlinear waves.

THE NONLINEAR SCHRÖDINGER EQUATION (N.L.S.)

An equation of the form (36) arises formally in the following way. Consider equation (19)

$$\nabla \times \nabla \times \mathbf{E} = \frac{\omega^2 \mathbf{E}}{c^2}\left[1 + \frac{\omega_{pe}^2}{\omega^2}\exp\left(-\frac{\varepsilon_0 |E|^2 \omega_{pe}^2}{4n_0 \omega^2 kT}\right)\right] \tag{41}$$

and assume $\mathbf{E}$ to be parallel to the y-direction.

Moreover we restore the time dependence of equation (41) that is we replace ω^2 by $-\partial^2/\partial t^2$ and assume $\partial/\partial x \equiv 0$, but leave ω^2 in the exponent.

We obtain the equation

$$\frac{d^2 E}{dy^2} + \frac{\partial^2 E}{\partial z^2} - \frac{1}{c^2}\frac{\partial^2 E}{\partial t^2} - \frac{\omega^2}{c^2}E\left[\exp\left(-\frac{\varepsilon_0 |E|^2(\omega_{pe}^2/\omega^2)}{4kTn_0}\right)\right] = 0. \tag{42}$$

Assuming E in the form

$$E = \mathscr{E}(z)\exp[i(\omega t - kz)] \tag{43}$$

or

$$E = \mathscr{E}(t)\exp[i(\omega t - kz)] \tag{44}$$

where $\mathscr{E}(z)$ is a slowly varying function of z we can neglect the second derivative of E and write

$$\frac{d^2\mathscr{E}}{dy^2} + \alpha^2\mathscr{E} - 2ik\frac{\partial\mathscr{E}}{\partial z} - \frac{\omega_{pe}^2\mathscr{E}}{c^2}\exp\left(-\frac{\varepsilon_0|\mathscr{E}|^2(\omega_{pe}^2/\omega^2)}{4kTn_0}\right) \qquad (45)$$

$$\alpha^2 = \left(\frac{\omega^2}{c^2} - k^2\right)$$

or, if we consider $\mathscr{E}$ as a slow varying function of t we can write

$$\frac{d^2\mathscr{E}}{dy^2} - \alpha^2\mathscr{E} + 2i\omega\frac{\partial}{\partial t} - \frac{\omega_{pe}^2}{c^2}\mathscr{E}\exp\left(-\frac{\varepsilon_0|\mathscr{E}|^2(\omega_{pe}^2/\omega^2)}{4kTn_0}\right) \qquad (46)$$

Now we consider the case of a weak quasi-potential, i.e.

$$\frac{\omega_{pe}^2}{\omega^2}\frac{\varepsilon|\mathscr{E}|^2}{2} \ll 2n_0kT \qquad (47)$$

and obtain, by expanding the exponential either

$$\frac{\partial^2\mathscr{E}}{\partial y^2} + \alpha^2\mathscr{E} - \frac{\omega_p^2}{c^2}\left(\frac{1+(\omega_{pe}^2/\omega^2)\varepsilon_0|\mathscr{E}|^2}{4TkTn_0}\right)\mathscr{E} - 2ik\frac{\partial\mathscr{E}}{\partial z} = 0 \qquad (48)$$

in case (43) or in case (44)

$$\frac{d^2\mathscr{E}}{dy^2} - \alpha^2\mathscr{E} - \frac{\omega_p^2}{c^2}\left(\frac{1-(\omega_p^2/\omega^2)\varepsilon_0|\mathscr{E}|^2}{4kTn_0}\right)\mathscr{E} + 2i\omega\frac{\partial\mathscr{E}}{\partial t} = 0. \qquad (49)$$

These equations are of the form

$$\psi_{xx} + k_1\psi + k|\psi|^2\psi \pm i\psi_z = 0 \qquad (50)$$

which is only trivially different from (36).

A SOLUTION OF THE N.L.S.

A solution of (36) will now be looked for in the form

$$\psi = \phi(x, t)\, e^{i\theta(x, t)} \qquad (51)$$

The exponential factors cancel and real and imaginary parts can be separated to obtain two equations

$$\phi_{xx} - \phi\theta_x^2 - \phi\theta_t + \beta\phi^3 = 0 \qquad (52)$$

$$\phi\theta_{xx} + 2\phi_x\theta_x + \phi_t = 0. \qquad (53)$$

We now seek a travelling wave solution with a carrier moving with velocity u_c and a phase given by

$$\theta = \theta(x - u_c t) = \theta(\xi) \tag{54}$$

modulated by an envelope ϕ travelling at a speed u_e in the form

$$\phi = \phi(x - u_e t) = \phi(\xi). \tag{55}$$

This assumption turns the partial differential equations (52, 53) into ordinary ones

$$\phi_{xx} - \phi\theta_x^2 + u_c\phi\theta_x + \beta\phi^3 = 0 \tag{56}$$

$$\phi\theta_{xx} + 2\phi_x\theta_x - u_e\phi_x = 0 \tag{57}$$

which can be integrated. Indeed, starting with (57) we find

$$\phi^2(2\theta_x - u_e) = \text{constant}. \tag{58}$$

It is convenient to assume that the constant is zero. Then $\theta_x = u_e/2$ and (56) integrates and yields

$$\frac{d\phi}{dx} = \left[-\frac{\beta}{2}\phi^4 + \tfrac{1}{4}(u_e^2 - 2u_e u_c)\phi^2 + C \right]^{1/2} \tag{59}$$

which is the equation of a mass point moving in a potential

$$P(\phi) = -\frac{\beta}{2}\phi^4 + \tfrac{1}{4}(u_e^2 - 2u_e u_c)\phi^2 + C. \tag{60}$$

It is clear from Fig. 7b showing this potential when $C = 0$ that, in this case, the motion of the point will have turning points at $\phi = \pm\phi_0$, and, having crossed the hollow will take a long (infinite) time to climb the hill. The polynomial has roots at $\phi = \pm\phi_0$ and mathematically speaking, the integral

$$\xi = \int_{\phi(0,\,0)}^{\phi(x,\,t)} \frac{d\phi}{(P(\phi))^{1/2}} \tag{61}$$

converges for small $\phi - \phi_0$, hence we have turning points. At $\phi = 0$ on the other hand, there is a double root and the integral is logarithmic, and becomes very large for small ϕ. The amplitude ϕ_0 is given by

$$\phi_0 = [(u_e^2 - 2u_c u_e)/2\beta]^{1/2} \tag{62}$$

and the solution is

$$\phi(x, t) = \phi_0 \operatorname{sech}\left[\left(\frac{\beta}{2}\right)^{1/2} \phi_0(x - u_e t) \right] \exp\left[i\frac{u_e}{2}(x - u_c t) \right] \tag{63}$$

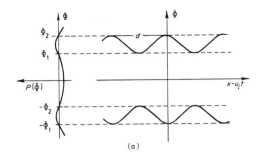

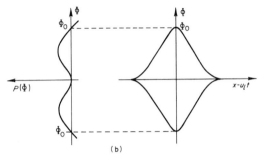

Fig. 7(a). Periodic envelope for solution of nonlinear Schrödinger equation.
Fig. 7(b) Soliton envelope for solution of nonlinear Schrödinger equation.

as may be found by carrying out the integration. When $C \neq 0$ the integral is found in terms of elliptic functions [4].

$$\phi(x, t) = \phi_1 \left[1 - \left\{ \left(1 - \frac{\phi_1^2}{\phi_2^2} \right) sn^2 \left[\left(\frac{\beta}{2} \right)^{1/2} (x - u_e t) \right] \right\} \right]^{1/2}$$

$$\times \exp \left[i \frac{u_e}{2} (x - u_c t) \right] \quad (64)$$

where sn has modulus $\gamma = 1 - \phi_1^2/\phi_2^2$ and the spatial period (wavelength)

$$d = \frac{4}{\phi_2} K(\gamma) \quad (65)$$

where K is the complete elliptic integral of the first kind. The solution in this more general case is illustrated in Fig. 7a, and the soliton solution in Fig. 7b.

The discussion parallels the previous one in the section on the quasi-potential where the exponential was not expanded. We now deal with transverse inhomogeneity. In the case of (43) the equation describes filamentation rather than confinement and the case (49) describes modulation.

E

For some equations which have soliton solutions one can also find multi-soliton solutions [4, 12] and this is the case for the KdV and the NLS equations, thus a multiplicity of filaments is described by such a solution in case (48).

It should be remarked that, interesting as our discussion may be, it is rather theoretical, because in practice, damping modifies the solutions in a rather obvious way which has been studied in, e.g. [12] of Chapter 7.

The KdV equation does not concern our main theme. Nevertheless it is interesting that it has another property, studied by Lax [9]. It has a spectrum of possible speeds of the soliton solutions and these speeds are the eigenvalues of an associated linear Schrödinger equation.

We have only studied one transverse direction (y) in the filamentation case. With one more dimension, one suspects "bubble" formation. The confinement problem also suggests the formation of radiation pockets.

Numerical studies by Estabrook et $al.$ [10] indeed show complicated structures evolving in the course of a time-dependent simulation treatment. It may be conjectured that, at high laser intensities, a kind of foam may be generated, with absorption and reflection properties which are hard to predict at this stage.

Simulation results are hard to interpret and we have presented detailed theoretical treatment of processes to supplement the understanding. We have studied elm waves, but similar behaviour was found with Langmuir waves by Zakharov [11] who also finds a non-NLS. Interesting speculations about the collapse of such solitons can be found in [14].

For the case of longitudinal oscillations of large amplitude, a type of collapse may arise which has been studied by Tang et $al.$ [15]. A self-consistent equation

$$(1 - 2\varepsilon^2)\,e^{-\varepsilon^2}\frac{d^2\varepsilon}{d\zeta^2} - (6 - 4\varepsilon^2)\,\varepsilon\,e^{-\varepsilon^2}\left(\frac{d\varepsilon}{d\zeta}\right)^2 + \beta(\beta - e^{-\varepsilon^2})\varepsilon = 0 \qquad (66)$$

for such oscillations was derived by these authors. It is here given in non-dimensional form with

$$\varepsilon = e|E|/4\omega(mT_e)^{1/2}, \qquad \beta = \frac{\omega^2}{m_p^2}, \qquad \zeta = (x/3)^{1/2}\,(T_e/m\omega_\nu^2)^{1/2}.$$

The solution of this equation diverges for $\varepsilon^2 > \frac{1}{2}$. This may be interpreted to mean that the high frequency radiative pressure is too strong to be balanced by the kinetic pressure of the surrounding electrons because of a substantial reduction in the number density and the wave collapses. A more exact equation given by Dorman [16] reads

$$(1 - \tfrac{2}{3}\varepsilon^2)\,e^{-\varepsilon^2}\frac{d^2\varepsilon}{d\zeta^2} - \tfrac{8}{3}\varepsilon\,e^{-\varepsilon^2}\left(\frac{d\varepsilon}{d\zeta}\right)^2 + \beta(\beta - e^{-\varepsilon^2})\varepsilon = 0. \qquad (67)$$

Its study leads to a similar conclusion for $\varepsilon^2 > \frac{3}{2}$. It may be, however, that the singularity arises as a consequence of the approximation procedure used in the derivation of (66) and (67).

In [15] it is also shown that the solutions of (66) and (67) have the same general features as those discussed in the case of transverse waves as illustrated by Figs 2 and 3. However, when ε^2 approaches $\frac{1}{2}$, the solutions of (66) become highly peaked and the same happens for $\varepsilon^2 \to \frac{3}{4}$ in the case of (67).

REFERENCES

1. Motz, H. and Watson, C. J. H. (1967). *Adv. Electron. Electron. Phys.* **23**.
2. Self, S. A. (1960). *Phys. Fluids,* 3, 488.
3. Drake, J., Kaw, P. K., Lee, Y. C., Schmidt, G. and Liu, C. S. and Rosenbluth, M. N. (1974). *Phys. Fluids,* **17**, 778–785.
4. Scott, A. C., Chu, F. Y. F. and McLaughlin, D. W. (1973). *Proc. IEEE,* **61**, 1443–1483.
5. Perring, J. K. and Skyrme, T. H. R. (1962). *Nucl. Phys.* **31**, 550–555.
   Skyrme, T. H. R. (1958). *Proc. Roy. Soc.* A **247**, 260–278.
6. Scott Russell, J. (1844). "Report on Waves" Report of the British Association.
7. Zabusky, N. J. and Krushal, M. D. (1965). *Phys. Rev. Lett.* **15**, 240.
8. Ikezi, H., Kiwamoto, Y. Nishikawa K. and Mima, K. (1972). *Phys. Fluids,* **15**, 1605.
9. Lax, P. D. (1968). *Comm. Pure Appl. Math.* **21**, 467–490.
10. Estabrook, K. (1975). *Phys. Fluids,* **19**, 1733–1738.
11. Zakharov, V. E. (1972). *Sov. Phys. JETP,* **35**, 908–914.
12. Zakharov, V. E. and Shabat, A. B. (1972). *Sov. Phys. JETP,* **34**, 62–69.
13. Hora, H. (1976). *Aust. J. Phys.* **29**, 375–388.
14. Nishikawa, K., Lee, Y. C. and Liu, C. S. (1975). *Comm. Plasma Phys.* **2**, 63.
15. Ting-Wei Teng and Motz, H. (1977). *IEEE Trans (Plasma Physics),* **4**, 297–300.
16. Dorman, G. (1969). *J. Plasma Phys.* **3**, 387.

7

Absorption of Laser Light and Plasma Radiation

In this chapter the emission and absorption of light by an inhomogeneous plasma surrounding the target will be studied. The problem is rather complex We start with the definitions of the quantities involved in radiation in a dielectric medium and its transfer from one position to another: energy density, spectral intensity, emission and absorption coefficients. We give expressions for the absorption coefficient in a homogeneous plasma and then proceed to discuss absorption in an inhomogeneous plasma. For laser energy density low compared to the thermal energy density linear absorption theory is appropriate. Free electrons cannot absorb light since energy and momentum cannot both be conserved in the elementary process. We shall therefore see that the collision frequency is crucially involved in the absorption process. Light emission occurs when the electrons are decelerated, one speaks of *Bremsstrahlung* (*bremsen* = slow down) or, in the case of absorption, of *inverse Bremsstrahlung*. When the laser power is large, the amount absorbed is no longer proportional to the laser intensity, linear theory breaks down because the parametric process, etc., discussed in a previous chapter set in. With still higher laser intensities we get filamentation, self modulation, etc. With increasing intensity the ponderomotive force gains in importance. Matters are, of course more complicated in the inhomogeneous medium. Large amplitude electric fields drive instabilities creating Langmuir oscillations with high field strength, particles are trapped in the fields and are ejected with high velocity. We soon reach the limits of analytic treatment and have to resort to computational methods to interpret experiments.

EMISSION AND ABSORPTION

We start by recalling Planck's distribution function for the number of light particles with wave vector k at some point r, $N(k, r)$ given by

$$N(\mathbf{k}, \mathbf{r}) - [\exp(\hbar\omega/kT) - 1]^{-1}. \tag{1}$$

The number of bosons $N(\mathbf{r})$ per unit volume of physical space is

$$N(\mathbf{r}) = \frac{1}{(2\pi)^3} \int N(\mathbf{k}, \mathbf{r}) \, d^3k \tag{2}$$

which comes from the fact that the number of waves with propagation vector $\mathbf{k}$ in volume element d^3k equals $(d^3k)\, V/(2\pi)^3$ where V is the volume. Now we can write

$$N(\mathbf{r}) = \frac{1}{8\pi^3} [\exp(\hbar\omega/kT) - 1]^{-1} \int d\omega\, k^2 \left|\left(\frac{\partial k}{\partial \omega}\right)\right| d\Omega_k \tag{3}$$

where $d\Omega_k = \sin\theta\, d\theta\, d\phi$ is the elementary solid angle about $\mathbf{k}$. We can introduce the refractive index $n_r = c/v_{ph} = ck/\omega$, and express the spectral energy density, (the product of $N(r)/d\omega$ and $\hbar\omega$) in a dispersive medium as

$$u_\omega = \frac{\hbar\omega^3}{8\pi^3 c^3} [\exp(\hbar\omega/kT) - 1]^{-1} \int n_r^2 \left|\left(\frac{\partial\omega n_r}{\partial\omega}\right)\right| d\Omega_k. \tag{4}$$

In vacuum we recover the more familiar expression and by means of $(1, 2)$

$$u_{0\omega} = \frac{\hbar\omega^3}{2\pi^2 c^3} [\exp(\hbar\omega/kT) - 1]^{-1} \tag{5}$$

while for an isotropic medium

$$u_\omega = u_0(\omega)\, n_r^2 \left|\left(\frac{\partial\omega n_r}{\partial\omega}\right)\right|. \tag{6}$$

For the treatment of an anisotropic medium see [5] of Chapter 4.

ENERGY DENSITY AND POWER FLUX

The energy which flows per unit time and unit area through a surface element at $\mathbf{r}$ is calculated by means of Poynting's vector to be

$$F(\mathbf{r}, t) = \mathbf{E}(\mathbf{r}, t) \times \mathbf{H}(\mathbf{r}, t) \tag{7}$$

and its time average for large time τ is defined by

$$\langle \mathbf{F}(\mathbf{r}, \tau) \rangle = \lim_{\tau \to \infty} \int_{-\tau/2}^{\tau/2} \frac{\mathbf{E}(\mathbf{r}, t, \tau) \times \mathbf{H}^*(\mathbf{r}, t, \tau) \, dt}{\tau}. \tag{8}$$

Quantities $Q(\mathbf{r}, t, \tau)$ are defined by

$$Q(\mathbf{r}, t, \tau) = Q(\mathbf{r}, t); \ \tau/2 \leqslant t \leqslant \tau$$
$$= 0; |t| > \tau/2$$

to ensure that Fourier transforms exist although $\int_{-\infty}^{\infty} |Q(\mathbf{r}, t)| \, dt$ does not. Taking Fourier transforms,

$$\mathbf{E}(\mathbf{r}, t, \tau) = \frac{1}{2\pi} \int_{-\infty}^{\infty} E(\mathbf{r}, \omega, \tau) \, e^{i\omega t} \, d\omega; \tag{9a}$$

$$\mathbf{H}^*(\mathbf{r}, t, \tau) = \frac{1}{2\pi} \int_{-\infty}^{\infty} H^*(\mathbf{r}, \omega', \tau) \, e^{-i\omega' t} \, d\omega' \tag{9b}$$

$$\mathbf{F}(\mathbf{r}, t, \tau) = \frac{1}{2\pi} \int_{-\infty}^{\infty} F(\mathbf{r}, \omega, \tau) \, d\omega; \tag{9c}$$

we find, since $\int_{-\infty}^{\infty} e^{i(\omega - \omega')t} \, dt = 2\pi\delta(\omega - \omega')$

$$F(r, \omega, \tau) = E(\mathbf{r}, \omega, \tau) \times H(\omega, \mathbf{r}, \tau). \tag{10}$$

The quantity

$$\langle F(r, \omega) \rangle = \lim_{\tau \to o} \frac{E(\mathbf{r}, \omega, \tau) \times H^*(\mathbf{r}, \omega, \tau)}{\tau} \tag{11}$$

is the average spectral energy flux density.

We now want to establish the connection between energy density and flux. Consider Fig. 1, the flux $d\mathbf{F}$ into a cone of opening $d\Omega$ in the direction of a ray will be denoted by $I_\omega(s) \, d\Omega$, when I_ω is measured in W cm^{-2} steradian. The energy flux through an element da of the area due to $I_\omega(s)$ is given by

$$dP(\omega) = d\mathbf{F}_\omega \cdot \mathbf{n} = I_\omega(s) \cos \theta \, d\Omega \, da \tag{12}$$

and the flux per unit area by

$$P(\omega) = \int I_\omega(s) \cos \theta \, d\Omega \tag{13}$$

where $d\Omega$ is the element of solid angle along the ray and $I_\omega(s)$ is the specific spectral intensity. For an observer at large distance in direction $\mathbf{n}$

$$P(\omega) = \int I_\omega \, d\Omega. \tag{14}$$

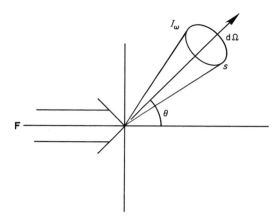

Fig. 1. Energy flux flowing into a cone of opening $d\Omega$.

Passing to the energy density, we express the time average by

$$\langle u(t) \rangle = \int_0^\infty u_\omega \, d\omega \tag{15}$$

where

$$u_\omega = \int_{4\pi} \frac{I_\omega}{v_{gr}} \, d\Omega \tag{16}$$

because energy propagates with the group velocity v_{gr} along the rays. *In vacuo* $v_{gr} = c$ and

$$cu_\omega = 4\pi I(\omega). \tag{17}$$

Now consider a ray passing from the vacuum into a medium m with refractive index n_r as illustrated in Fig. 2. By Snell's law, $n_r \sin \theta_m = \sin \theta$ and for small angle $\Delta\theta$

$$\sin \theta \, d\theta = (\Delta\theta)^2, \qquad d\Omega_0/d\Omega_m = (\Delta\theta_m/\Delta\theta_0)^2 = n_r^2.$$

Since

$$I_0 \, d\Omega_0 = I_m \, d\Omega_m$$

$$I_m(\omega) = n_r^2 I_0(\omega) \tag{18}$$

and so

$$\frac{I_m(\omega)}{n_r^2} = \text{constant} \tag{19}$$

along the ray. Similarly from (16)   (20)

$$u_\omega = n_r^2 u_{0\omega}.$$

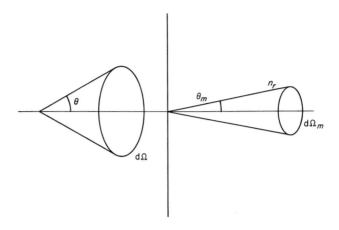

Fig. 2. Ray passing from vacuum into a refracting medium.

EMISSION AND ABSORPTION COEFFICIENTS

When there is light emission in a medium, we define an emission coefficient j_ω as the power emitted per unit volume per unit $d\omega$ and unit solid angle.

The absorption coefficient $\alpha(\omega)$ is defined by the absorption law

$$dI_\omega = -\alpha_\omega I_\omega \, ds. \tag{21}$$

From conservation of energy, taking (19) into account it is easy to deduce the radiation transport equation

$$\frac{d}{ds}\left(\frac{I_\omega}{n_r^2}\right) = \frac{j_\omega}{n_r^2} + \alpha_\omega \frac{I_\omega}{n_r^2}. \tag{22}$$

The quantity $S_\omega = j_\omega/\alpha_a n_r^2$ is known as the source function, and another quantity defined by

$$\tau = \int_0^\tau d\tau = -\int_0^s \alpha_\omega \, ds \tag{23}$$

is the optical depth. In terms of these quantities

$$\frac{d}{d\tau}\left(\frac{I_\omega}{n_r^2}\right) = \frac{I_\omega}{n_r^2} - S_\omega \tag{24}$$

which, in integrated form reads

$$I_\omega = I_\omega(\text{inc}) \, e^{-\tau_0} + \int_0^{\tau_0} S_\omega(\tau) \, e^{-\tau} \, d\tau \tag{25}$$

where $I_\omega(\text{inc})$ is the incident intensity. It expresses the intensity emerging from a ray passing through a medium with radiation sources when the total optical depth, i.e. the integral of (23) along the optical path is given by τ_0.

BLACK BODY RADIATION

The intensity of black body radiation will be referred to as $I_{\omega b} = B(\omega, T)$ where T is now the temperature. With the relation between specific intensity and energy density we obtain from (5)

$$B_0(\omega, T) = \frac{\hbar\omega^3}{8\pi^3 c^2} [\exp(\hbar\omega/kT) - 1]^{-1} \tag{26}$$

in vacuo. In a medium, taking account of (28) we obtain

$$B_m = n_r^2 B_0 \tag{27}$$

for the energy density.

Integrating (26) over ω and Ω one obtains the Stefan Boltzmann law

$$\text{Flux} = 5 \cdot 67 \cdot 10^{-8} \, T^4 \, \text{W m}^{-2} \tag{28}$$

valid for an optically thick medium. For a medium of thickness L one obtains

$$I_\omega = \frac{j_\omega}{n_r^2 \alpha_\omega} [1 + e^{-\alpha_\omega L}] \tag{29}$$

for a homogeneous medium by integrating (22). In thermal equilibrium, emission is balanced by absorption so that

$$j_\omega = \alpha_\omega B_m(\omega, T). \tag{30}$$

Expanding the exponential in (29) we see that the radiation intensity from a plasma slab of thickness L will be given by

$$I_\omega = \alpha_\omega L B_0 \tag{31}$$

i.e. it will be smaller than the black body radiation by a factor αL or L/λ_{abs} where the absorption length λ_{abs} is defined as the inverse of the absorption coefficient. We shall see that λ_{abs} is the order of 1 m for our laser plasma so that the factor is very small indeed. Thus a plasma would have to be very thick indeed to emit black body radiation according to formula (28).

THE ABSORPTION COEFFICIENT

We now give a simple account of the determination of the linear absorption coefficient, deferring the discussion of nonlinear effects. Consider a Fourier component $\exp(i\omega t + i\mathbf{k} \cdot \mathbf{r})$ or rather its square since we deal with power.

The absorption coefficient is given by

$$\alpha_m = -2\,\mathrm{Im}\,k \tag{32}$$

and to find α_m all we have to do is to compute the dispersion relation, which, for elm. waves is given by (47) of Chapter 4 as

$$k^2 c^2 = \omega^2 \varepsilon_{11} (\varepsilon_{11} = \varepsilon_{22}) \tag{33}$$

where ε_{11} is given by (32) as $1 + \mathbf{T}_{11}/i\varepsilon_0 m$ of the medium and $\mathbf{T}_{11} = \omega_p^2 \varepsilon_0 S^{-1}$. In the presence of collisions $\mathbf{S}_e^{-1}$ is given by (80) where v_{en} is the electron atom or, in the case of a fully ionized laser fusion plasma, is the electron ion collision frequency v. Splitting $(1 + v/i\omega)^{-1}$ into real and imaginary parts

$$\varepsilon_{11} = 1 - \frac{\omega_p^2}{\omega^2 + v^2} - i\frac{\omega_p^2}{\omega^2 + v^2}\frac{v}{\omega}. \tag{34}$$

From (32) and (33) we find, since $\mathrm{Im}\,k^2 = 2\,\mathrm{Im}\,k\,\mathrm{Re}\,k$

$$\alpha_m = -\frac{\omega}{c}\frac{\mathrm{Im}}{n_r}\varepsilon_{11} \tag{35}$$

where n_r is the refractive index

$$n_r = \frac{c}{\omega}\,\mathrm{Re}\,k \tag{36}$$

We therefore obtain

$$\alpha_\omega = \frac{v}{cn_r}\frac{\omega_p^2}{\omega^2 + v^2} \approx \frac{v}{cn_r}\frac{\omega_p^2}{\omega^2} \quad \text{for} \quad v \ll \omega. \tag{37}$$

Note, that $n_r \approx (1 - \omega_p^2/\omega^2)^{1/2}$ (for $v \ll \omega$) and that the absorption becomes very large as $\omega \to \omega_p$. The emission coefficient j_ω will be given by (30) as

$$j_\omega = \alpha_\omega B = \alpha_\omega n_r^2 \frac{\hbar\omega^3}{8\pi^3 c^2}\left[\exp(\hbar\omega/kT) - 1\right]^{-1}. \tag{38}$$

The collision frequency averaged over a Maxwellian distribution is given by [5, p. 99] of Chapter 4

$$\langle v \rangle = \frac{\sqrt{3}}{9\pi^{\frac{1}{2}}}\left(\frac{Ze^2}{m\varepsilon_0}\right)\left(\frac{m_e}{2kT}\right)^{3/2}\omega_p^2\bar{G}(T,\omega) = 2{\cdot}07\,.\,10^{-9}\frac{Z\omega_p^2}{T^{3/2}}\bar{G} \tag{39}$$

where $\bar{G}$, the "Gaunt factor", arising from quantum corrections, is

$$\bar{G} = \frac{\sqrt{3}}{\pi}[19\cdot56 + \ln(T^{3/2}/\omega Z)] \quad \text{for} \quad T \leqslant 10^6 \text{ K};$$

$$\bar{G} = \frac{\sqrt{3}}{\pi}[26\cdot41 + \ln(T/\omega)] \quad \text{for} \quad T \geqslant 10^6 \text{ K}.$$

(40)

Dawson and Oberman [1] have a somewhat different approach. Referring back to expression (42) of Chapter 4, in the absence of a magnetic field the tensor becomes a scalar and we can simply write

$$\varepsilon = 1 - i/\varepsilon_0 \omega Z(\omega)$$

(41)

where $Z(\omega)$ is the high frequency resistance. These authors then proceed to find this quantity by means of a Vlasov equation treatment and get

$$Z(\omega) \approx \frac{i\omega}{\varepsilon_0 \omega_p^2}\left[1 - (2\pi)^{3/2}\frac{Ze^2\omega_p^2}{6mv_T^3\omega}\right.$$
$$\left. - i\left(\frac{\pi}{2}\right)^{-1/2}\frac{Ze\omega_p^2}{6mv_T^3\omega}\left[\ln\left(\frac{2k_{max}^2 v_T^2}{m\omega^2}\right) - 0\cdot577\right]\right]$$

(42)

where the value to be used for k_{max} which is needed as a cut-off to avoid divergence of an integral is somewhat in doubt. The best available absorption coefficient is due to Johnson and Dawson [2] and is given by

$$\alpha_0 = \frac{n_e n_i Z^2 e^6}{6\pi\mu\varepsilon_0^3 c\omega^2}\left(\frac{1}{2\pi}\right)^{1/2}\left(\frac{1}{m_e k T_e}\right)^{3/2}\frac{\pi}{\sqrt{3}}\bar{g},$$

(43)

with $\bar{g} = \sqrt{3}/\pi \ln \Lambda$, where Λ is taken to be the smaller of $v_T/\omega_{pe}p_{min}$ and $v_T/\omega p_{min}$ and p_{min}, the minimum of the electron impact parameter, is the larger of $Ze^2/4\pi\varepsilon_0 k T_e$ and $\hbar/(m_e k T_e)^{1/2}$.

Details are hardly important, however, as we still have to take account of modifications due to the high laser intensity, nonlinear effects, and the effect of inhomogeneity. Firstly, we remark that for high laser intensity the thermal velocity $\sqrt{kT/m_e}$ should presumably be replaced by the quiver velocity $eE/m_e\omega$ in the laser field. Secondly, it has been argued [3] that the collision frequency in formula (37) might be replaced by a suitably determined turbulent collision frequency.

We can discuss the nonlinear development in terms of the ratio η of the quiver velocity v_E and the thermal velocity v_T

$$\eta_E = \frac{v_E}{v_T}.$$

(44)

In terms of this parameter, the threshold for the OTS for $\eta \ll 1$ is given by

$$\eta^2 = 4\left(\frac{v_e}{\omega_{pe}}\right)\left(1 + \frac{T_i}{T_e}\right).$$ (45)

Kaw and Dawson [3] argue that, at the threshold field, strong ion fluctuations appear and that, as a result the effective collision frequency increases until at a value of $v_{e\,max}$ the laser field is no longer above threshold. This will happen when, according to (45)

$$v_{emax} \approx \eta^2 \omega_{pe}/4.$$ (46)

Substituting this value in (37) the maximum absorption coefficient for this instability is given by

$$\alpha_{max} \approx \frac{30I_\omega}{n_e^{1/2}T_e}$$ (47)

where I_ω is in W cm^{-2} and T_e in eV. Near ω_p in the range $0.9 \geqslant \omega_{pe}/\omega \geqslant 0.7$, α_0 may be written

$$\alpha_0 \approx 2.10^{-15}\frac{n_e}{T_e^{3/2}}\text{ cm}^{-1}$$ (48)

so that the parametric absorption exceeds the *Bremsstrahlung* absorption when

$$I_\omega \geqslant 10^{-16}\frac{n_e}{T_e^{1/2}}\text{ W cm}^{-2}.$$ (49)

For large values of η the wave field becomes large enough to trap electrons. Computer simulations by various workers, De Groot and Katz, [4] and by Kruer and Dawson [6] etc. lead to interesting conclusions which we cannot discuss in detail. The book by Hughes, [5], an invaluable sourcebook for the subject of laser plasma interaction, particularly in the homogeneous case, contains a good exposition.

ABSORPTION IN THE INHOMOGENEOUS PLASMA

In theoretical discussions of absorption in an inhomogeneous plasma it is assumed that the density gradient is either linear or quadratic. We briefly assemble some mathematical tools for dealing with these situations.

The case of a linear density gradient leads to a differential equation

$$\frac{d^2E}{d\zeta^2} = \zeta E$$ (50)

which is known as the Stokes equation. In terms of series of ascending powers the solution is given by

$$E = a_0 \left\{ 1 + \frac{\zeta^3}{3 \cdot 2} + \frac{\zeta^6}{6 \cdot 5 \cdot 3 \cdot 2} + \frac{\zeta^9}{9 \cdot 8 \cdot 6 \cdot 5 \cdot 3 \cdot 2} + \cdots \right\} \qquad (51)$$

$$+ a_1 \left\{ \zeta + \frac{\zeta^4}{4 \cdot 3} + \frac{\zeta^7}{7 \cdot 6 \cdot 4 \cdot 3} + \frac{\zeta^{10}}{10 \cdot 9 \cdot 7 \cdot 6 \cdot 4 \cdot 3} \right\}.$$

There are two independent solutions $Ai(\zeta)$ and $Bi(\zeta)$, and for $Ai(\zeta)$, $a_0 = 3^{-2/3}/(-\frac{1}{3})!$, $a_1 = (-3)^{-1/3}/(-\frac{2}{3})!$; while for $Bi(\zeta)$, $a_0 = (3)^{-1/6}/(-\frac{1}{3})!$, $a_1 = 3^{1/6}/(-\frac{2}{3})!$ By themselves the two series have no particular significance. We shall be particularly interested in the function $Ai(\zeta)$ known as the Airy integral which may be written in the form

$$Ai(\zeta) = \frac{1}{\pi} \int_0^\infty \cos(\zeta s + \tfrac{1}{2}s^3) \, ds \qquad (52)$$

and has the general form shown in Fig. 3. When $\zeta \gg 1$, $E = \zeta^{-1/4} \exp(\pm \tfrac{2}{3}\zeta^{3/2})$ are good approximations to solutions of (50), as may be proved from (52) by the method of steepest descent.

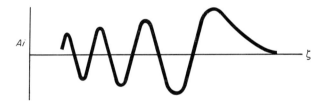

Fig. 3. Shape of the Airy function Ai (ζ).

The subject of electromagnetic waves in inhomogeneous media has been much studied in connection with ionospheric radio wave propagation and a very full account is given in the book by Budden [7]. The case of the parabolic approximation to the density distribution also leads to an analytically soluble equation, Weber's equation, which is again studied in [7].

WAVE BEHAVIOUR IN AN INHOMOGENEOUS PLASMA

From Maxwell's equations

$$\nabla \times \mathbf{E} = -i\omega\mathbf{B}$$

$$\nabla \times \mathbf{H} = -n_e(x)e\mathbf{v} + i\omega\varepsilon_0\mathbf{E}$$

one obtains the linear wave equation

$$\nabla \cdot (\nabla \cdot \mathbf{E}) - \nabla^2 \mathbf{E} - \frac{\omega^2}{c^2} \mathbf{E} = i\omega e n_0(x)\mathbf{v}/\varepsilon_0 c^2. \tag{53}$$

In linear theory we write

$$\frac{\partial \mathbf{v}}{\partial t} = -\frac{e}{m}\mathbf{E} - \nu\mathbf{v} \tag{54}$$

where ν is a collision frequency and we have assumed

$$\mathbf{E} = \mathbf{E}\exp\left[i(\omega + \mathbf{k}\cdot\mathbf{r})\right]. \tag{55}$$

We first treat the case of normal incidence of the wave on a plasma with a linear gradient in the x-direction illustrated by Fig. 4. We have $k_y = k_z = 0$

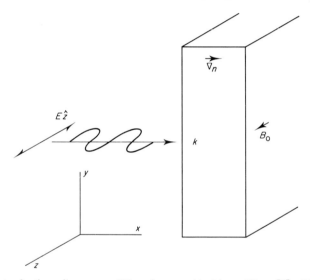

Fig. 4. Geometry for the ordinary wave, TE mode, normal incidence. (From [8], with permission.)

and assume polarization in the z-direction. Neglecting collisions we obtain the equation

$$\frac{d^2 E_z}{dx^2} + k_0^2\left(1 - \frac{\omega_p^2}{\omega^2}\right)E_z = k_0^2\varepsilon E_z = 0 \tag{56}$$

for the amplitude. It is now assumed that the density varies linearly as illustrated in Fig. 5 so that ε which is the index of refraction is given by

$$\varepsilon = 1 - \frac{\omega_p^2}{\omega^2} = 1 - \left(\frac{x}{x_0}\right). \tag{57}$$

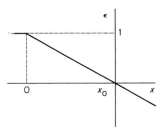

Fig. 5. Linear behaviour of the dielectric function giving rise to Airy function solution.

Making the transformation $\zeta = (k_0^2/x_0)^{1/3}(x_0 - x)$ brings (56) into the form

$$\frac{d^2E}{d\zeta^2} + \zeta E = 0 \tag{58}$$

which has $Ai(-\zeta)$ as a solution. It is illustrated in Fig. 6 for the case $k_0 x_0 = 70$.

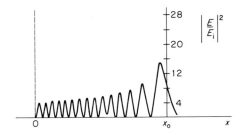

Fig. 6. Wave intensity distribution for linear profile, $k_0 x_0 = 70$. (From [8], with permission.)

The electric field strength reaches a maximum several wavelengths before the cut-off layer. The boundary conditions lead to the solution

$$E(-\zeta) = A \int_0^\infty \cos\left(\tfrac{1}{3}x^3 - \zeta x\right) dx \tag{59}$$

and the maximum is given by

$$\left|\frac{E_m}{A}\right|^2 = 0\cdot935 \tag{60}$$

which, at the boundary where the distance $\zeta_0 = (k_0^2/x_0)^{1/3}x_0 = (k_0 x_0)^{2/3}$ is sufficiently large may be approximated by

$$E \approx A\zeta_0^{-1/4} \cos\left(\tfrac{2}{3}\zeta^{3/2} - \pi/4\right) \tag{61}$$

at a peak near the boundary

$$|E/A|^2 = (k_0 x_0)^{-1/3}. \tag{62}$$

This peak results from the constructive interference of reflected and incident waves of the same amplitude hence for the incident wave $(E_i/A)^2 = (k_0 x_0)^{-1/3}/4$. Dividing 0·935 by this gives the amplification factor

$$|E_m/E_i|^2 = 3·74 \cdot (k_0 x_0)^{1/3}. \tag{63}$$

If a quadratic density profile is assumed the result is higher by 47%. When the collision frequency is taken into account ε is given by

$$\varepsilon = 1 - \frac{\omega_p^2}{\omega^2}(1 + iv/\omega)^{-1}. \tag{64}$$

The wave is attenuated on the way in and out and amplification factor is reduced by a factor $\exp[-\frac{4}{3}(v/\omega)k_0 x_0]$.

The case of oblique incidence has rather different features according to whether the polarization is in the plane of incidence as in Fig. 7 or perpendicu-

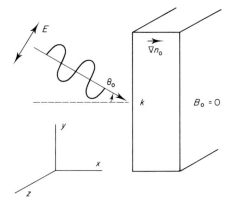

Fig. 7. Geometry for ordinary wave, TM mode, oblique incidence. (From [8], with permission.)

lar to it, as in Fig. 8. First we deal with the case of Fig. 8. From (53) we now obtain

$$\frac{d^2 E_z}{dx^2} - k_y^2 E_z + k_0^2\left(1 - \frac{\omega_p^2}{\omega^2}\right) = 0. \tag{65}$$

Since $k_y = k \sin \theta$ is constant by Snell's law it is equal to its value $k_0 \sin \theta_0$ in the uniform region so that

$$\frac{d^2 E_z}{dx^2} + k_0^2 \varepsilon' E_z = 0, \qquad \varepsilon' = \left(1 - \frac{\omega_p^2}{\omega} - \sin^2 \theta_0\right). \tag{66}$$

The cut-off now occurs at a lower value of ω_p given by

$$\omega_p = \omega \cos \theta_0. \tag{67}$$

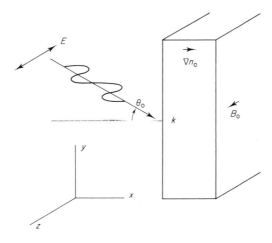

Fig. 8. Geometry for the ordinary wave, TE mode, oblique incidence.

For a linear profile substituting $\zeta \equiv (k_0^2/x_0)^{1/3} (x_0 \cos^2 \theta_0 - x)$ leads again to equation (58) and therefore the intensity variation is still as in Fig. 6 except for the position of the cut-off now occurs at the point

$$x_c = x_0 \sin^2 \theta_0 \tag{68}$$

The case of polarization in the plane of incidence, as illustrated in Fig. 7 shows a very different behaviour. As the electrons oscillate in the field, as shown in Fig. 9, a component of their motion lies in the direction of the

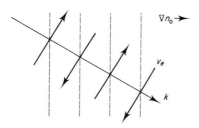

Fig. 9. Motion of electrons in the ordinary wave, TM mode, oblique incidence. (From [8], with permission.)

gradient of n_0. This causes charge separation and the wave cannot remain purely electromagnetic. The resulting field distribution is shown in Fig. 10.

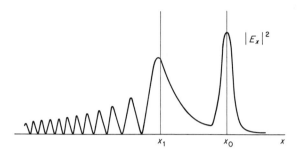

Fig. 10. Field distribution in the ordinary wave, TM mode, oblique incidence. (From [8], with permission.)

There is field enhancement near the point of reflection x_1 then the field becomes evanescent but excites plasma oscillations at the resonance layer where $\varepsilon = 0$. Details of this behaviour are discussed by White and Chen [8] and numerical computations have been carried out by Friedberg et al. [10]. The linear profile has been treated extensively by Denisov [9] and Fig. 11 which shows the dependence of the amplification factor on the angle is taken

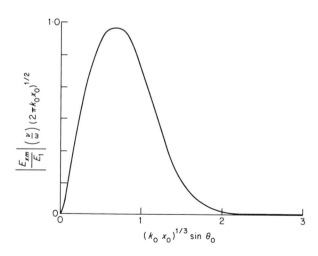

Fig. 11. Amplification factor as a function of $\tau = (k_0 x_0)^{1/3} \sin \theta_0$ for the ordinary wave. TM mode. (From [9], with permission.)

from his paper. The problem is however complicated by the effect of the ponderomotive force which becomes prominent at high fields, as shown by Hora ([20] of Chapter 8). Freidberg *et al* [10] compute values as high as 50% for the absorption coefficient in the angular range of maximum amplification and computer simulation by Kindel *et al.* ([7] of Chapter 12) lead to values even higher, of 90%.

The optimum angle for such absorption may be found by the following consideration. The distance between the resonant position and cut-off is, from (68)

$$x_0 - x_1 = x_0 \sin^2 \theta_0 \qquad (69)$$

and the field near the resonant layer will be of the form

$$E_r = E_c \exp(-k_0 x_0 \sin^2 \theta_0) \qquad (70)$$

where E_c is the field component at the critical layer and is given by

$$E_c = E_0 \sin \theta_0. \qquad (71)$$

Thus the resonant field amplitude will have the form

$$E_r(\theta_0) \propto E_0 \sin \theta_0 \exp(-k_0 x_0 \sin^2 \theta_0). \qquad (72)$$

The maximum field enhancement will occur at an angle obtained by setting the derivative of (72) with respect to θ_0 equal to zero which yields

$$(2k_0 x_0)^{1/2} \sin \theta_0 = 1. \qquad (73)$$

A more complete treatment by Ginzburg [11] yields the result

$$E_c(\theta_c) = E_0 \phi(\tau) \omega_0 / (2\pi k_0 x_0)^{1/2} v_e \qquad (74)$$

in terms of a parameter $\phi(\tau)$ involving Airy functions. The optimum angle is given (as in Fig. 11) by

$$\tau = (k_0 x_0)^{1/3} \sin \theta_0 = 0.8 \qquad (75)$$

and the resonant field amplitude by

$$E_{max} = \frac{1.2}{(2\pi k_0 x_0)^{1/2}} \frac{\omega_0}{v_e} E_0. \qquad (76)$$

The enhancement is high when $\omega_0 \gg v_e$. It leads to a high absorption which is of great importance for laser fusion (see Chapters 8 and 13). So far we have treated linear absorption, and it is seen from (76) that $E_{max}^2 \propto \omega_0$. It is modified by nonlinear effects discussed below.

Nonlinear Absorption

We now discuss the phenomena which occur for larger laser intensity in the inhomogeneous plasma, when the effect of the ponderomotive, or non-linear, force come more and more into play. Hora ([20] of Chapter 8) distinguishes this range as the one which occurs when the electromagnetic energy in the plasma in the resonance region exceed the vacuum value by more than the thermal energy and computes values

$$I_v > 7\cdot5 \,.\, 10^{10} T^{1/4} \,\text{W cm}^{-2} \quad \text{for the } CO_2 \text{ laser and}$$

$$I_v > 7\cdot5 \,.\, 10^{13} \, T^{1/4} \,\text{W cm}^{-2} \quad \text{for the neodynium laser } (T \text{ in eV}).$$

Going back to the normal incidence, we first assume the plasma to be driven by an external capacitor field E_0 which creates a longitudinal self consistent field E with time dependence slow compared with the laser period $2\pi/\omega_0$. The theory of the previous section describes linear response or linear conversion i.e. enhancement due to the Airy pattern. Now we consider a generalized nonlinear Schrödinger equation of the type discussed in the section on solitons. Morales and Lee [12] obtain it in the form

$$i \frac{2}{m_0} \frac{\partial}{\partial t} E + k_D^{-2} \frac{\partial^2}{\partial x^2} E + \varepsilon E = E_0 \tag{77}$$

where $k_D = \omega_0/\sqrt{3}v_T$ and v_T is the electron thermal velocity. The effect of the linear density gradient (Fig. 5) is taken into account in the following way:

When the laser field is high, the enhanced field expels plasma as explained in our section on the ponderomotive force in the case of transverse waves. Let the density change be δn. The dielectric coefficient ε is now given by

$$\varepsilon = -\frac{x}{x_0} + i \frac{v}{\omega_0} + \frac{\delta n}{n_0} \tag{78}$$

where v is again the collision frequency and n_0 the density at $x = 0$. δn is obtained by equating the particle pressure to the nonlinear force and is given by

$$\delta n/n_0 = -\varepsilon_0 |E^2|/2n_0(T_e + T_i). \tag{79}$$

In terms of the non-dimensional parameters defined by

$$\tau = \omega_0 t/(2k_D x_0)^{2/3}, \qquad A = E/E_0(k_D x_0)^{2/3}$$

$$z = (k_D^2/x_0)^{1/3} x, \qquad \Gamma = (v/\omega_e)(k_D x_0)^{2/3} \tag{80}$$

equation (77) reads:

$$i\frac{\partial A}{d\tau} + \frac{\partial^2 A}{dz^2} - (z - i\Gamma - p|A|^2)A = 1 \qquad (81)$$

where the parameter $p = (k_D x_0)^2 \varepsilon_0 E^2 / 2n_0(T_e + T_i)$ characterizes the strength of the non-linearity.

For small p the time dependence of $|A|^2$ is illustrated in Fig. 12 for times in the range $0.5 \leqslant \tau \leqslant 3.0$ and Γ is 0.2. Here $\mathrm{Re}(\varepsilon)$ is simply $-z$. The cold plasma resonance is at $z = 0$. When the driving field is suddenly turned on at $\tau = 0$ the cold plasma resonance is excited. As time progresses convection becomes important and the peak of $|A|^2$ is shifted towards $z < 0$. The Airy pattern begins to develop an outgoing wave which propagates down the

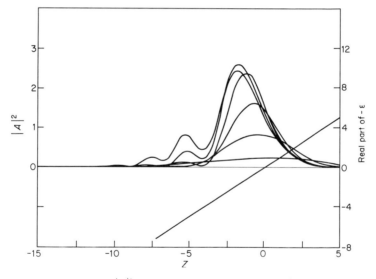

Fig. 12. Spatial dependence of $|A|^2$ and Re ε for $p = 0$, $\Gamma = 0.2$, in the interval $0.5 \leqslant v \leqslant 3.0$. (From [12], with permission.)

density gradient. For transverse waves this is the situation which obtains in the case of oblique incidence, when there is a field component in the direction of the density gradient. When the pump strength is in the range $0.1 \leqslant p \leqslant 0.8$ the Airy pattern is distorted by the nonlinear force. First the resonance point is shifted as in Fig. 12. A further distortion caused by the ponderomotive force at a later time ($\tau = 2.75$) is seen in Fig. 13. The case $p = 0$ is included for comparison. The figure also shows the real part of ε for $p = 0$ and $p = 0.5$.

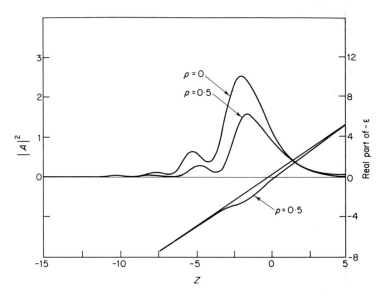

Fig. 13. Spatial dependence of $|A|^2$ and Re ε at $\tau = 2\cdot75$, $\Gamma = 0\cdot2$, for $p = 0, 0\cdot5$. (From [12], with permission.)

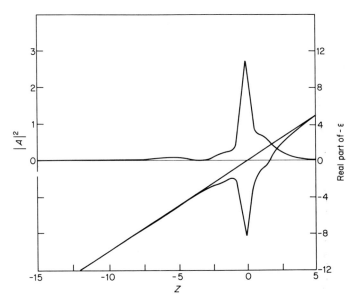

Fig. 14. Spatially localized field seen at large amplitude, $p = 3\cdot0$, $\Gamma = 0\cdot2$, $\tau = 1\cdot75$. (From [12], with permission.)

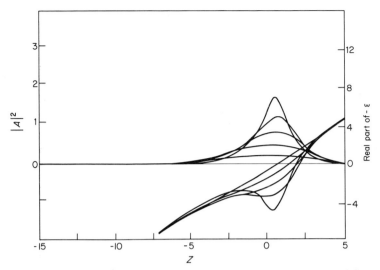

Fig. 15. Time evolution of $|A|^2$ for a large-amplitude case, $p = 3\cdot0$, $\Gamma = 0\cdot2$. (a) $0\cdot5 \leqslant \tau \leqslant 1\cdot5$. (From [12], with permission.)

For higher pump strength $p > 1$, an important new feature appears. The trapping of the field in the resonance region produces a dip in the plasma density which shifts the resonant point and a new resonance developed which can be seen as a shoulder on the field curve of Fig. 14 which has been drawn

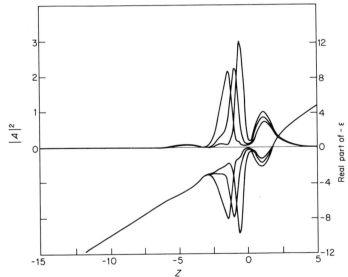

Fig. 16. Further time evolution of $|A|^2$ and $\mathrm{Re}(\varepsilon)$ in the large amplitude case, $p = 3\cdot0$, $\Gamma = 0\cdot2$, $2 \leqslant \tau \leqslant 2\cdot2$. (From [12], with permission.)

for $T = 1.75$. Figure 15 illustrates the behaviour in the time range $0.5 \leqslant \tau$ <1.5. As the field builds up, the nonlinear (ponderomotive) force begins to dig a cavity centred at $z = 0$. We have already shown the spiky cavity which develops thereafter at $\tau = 1.75$. At a later time the cavity breaks up into two isolated components illustrated in Fig. 16, for $2.0 \leqslant \tau \leqslant 2.2$, each of which develops its own density cavity. These cavities have been called cavitons, as they are relatively stable nonlinear configurations, analogous to solitons. Their existence has been substantiated by microwave experiments of Kim *et al* [12]. A more complete theory must also take into account coupling with ion sound and the effect of transverse waves. See [24, pp. 202, 211] and [25].

MICROWAVE SIMULATIONS

There are several areas of experimentation where the theories developed in preceding sections are applicable: Microwave experiments, explorations of the ionosphere and laser experiments. The corresponding scales of frequency, length scale and applied external fields are displayed in Fig. 17.

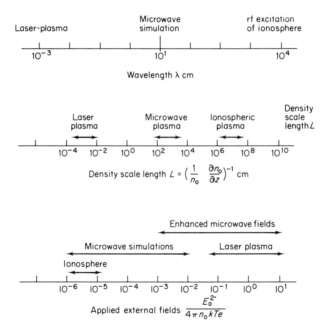

Fig. 17. A comparison of the experimental parameters, wavelength λ, density scale length L, and applied external fields E_0^2 for the three cases of interest—laser–plasma interaction, microwave simulation, and r-f excitation of the ionosphere. The microwave parameter regime overlaps the laser and ionospheric regimes in L/λ (number of free space wavelength in a density gradient length) and the normalized field strength $E^2/4\pi n_0 T$ where E is the self-consistent field in the plasma.

In the microwave experiments of the UCLA group a quiescent plasma is formed by ionizing gas with heated filaments in a large cylindrical vessel. Waves are launched by suitable aerials and it is possible to maintain an approximately linear gradient of plasma density along the cylinder axis. The experimental details may be found in a review by A. Y. Wong [14]. The observations show a large enhancement of the field in the case of oblique incidence when E has a component parallel to the density gradient. Large ion currents with ion energy much larger than the ambient develop just outside the resonance region. This is followed by a density depression of 40 % as a result of the expulsion of the ions from the resonant region. As this caviton structure travels down the density gradients the peak consisting of ions with a wide range of velocities quickly disperses leaving the density caviton in existence for a much longer time. All this may be seen in Fig. 18

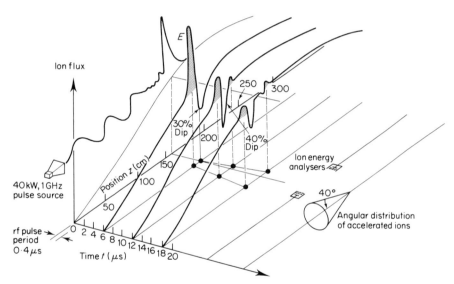

Fig. 18. Space time representation of ion bursts (shaded) driven by the ponderomotive force. Density cavities are created as a result of ion expulsion. The polarization of ion trajectories is represented by the cone with a half width of 40° with respect to the z-axis. (From [14], with permission.)

which shows the field pattern and the (shaded) ion pulses and cavitons as they evolve as a function of time. The angular distribution of accelerated ions is indicated by the 40° cone. High energy ions are always observed in laser irradiation of targets. The origin of these ions is not certain, but the microwave evidence points to nonlinear force effects.

High energy electrons are also found in these experiments. They can arise from electrons with speed larger than thermal (e.g. as produced by OTS) being accelerated in their transit through cavitons. If they traverse field regions in a time τ shorter than half the oscillation period the $2\pi/\omega_0$ energy gained will be of the order

$$eE \cdot l = eE\frac{\tau}{v} = \frac{eE\pi}{\omega_0} /(kT/m)^{1/2} = E\pi\sqrt{\varepsilon_0}/(kTn)^{1/2}m. \qquad (82)$$

Electrons leaving the field region which are responsible for the development of the caviton, on the other hand, would be expected to have energies equal to the quasi-potential which may be very high. Such electrons might penetrate the target and heat it prior to compression, thus increasing the pressure which resists the desired compression as discussed in Chapter 2.

The high energy ion flux will also diminish the energy available for compression. Ion energies in the kilovolt range have been observed in Laser fusion experiments which may produce neutrons by collisions with deuterons and tritons which are not of thermonuclear origin.

Spontaneous Generation of Magnetic Fields

Magnetic fields generated in the plasma as a result of electromagnetic radiation will significantly affect its transport properties as we shall see below. A theoretical treatment which predicts both thermoelectric currents and currents due to electron motion in the resonant region has been given by Thomson, Max and Estabrook [15]. They consider the geometry of incidence illustrated in Fig. 7. and use the momentum equation. They assume that the term $m\langle n\mathbf{vv}\rangle$ representing the electron inertia tensor due to drift motion is balanced by a ponderomotive stress tensor (discussed in [1] of Chapter 6) according to

$$m\nabla \cdot n(\mathbf{vv}) = \nabla \cdot [\varepsilon_0(1 - \varepsilon)\langle \mathbf{EE}\rangle]. \qquad (83)$$

They calculate the curl of the average electric field as calculated from the momentum equation and find the time change of the magnetic field to be given by the sum of two terms

$$\frac{\partial\langle B_T\rangle}{\partial t} = \frac{k}{n_0 e}\nabla T \times \nabla n_0 \qquad (84)$$

and

$$\frac{\partial\langle B_z\rangle}{\partial t} = \frac{\partial}{\partial x}\left\{\frac{\varepsilon_0}{en_0}\frac{\partial}{\partial x}\left[\frac{\omega_p^2}{\omega^2}\langle E_x E_y\rangle\right]\right\} = \frac{\partial}{\partial x}\left\{\frac{\varepsilon_0}{en_0}\frac{\partial}{\partial x}\left[\frac{x}{x_0}\langle E_x E_y\rangle\right]\right\} \qquad (85)$$

where the last equation holds for the density variation (57). The first results from thermoelectric effects and the second is the electromagnetic term. It involves the average $\langle E_x E_y \rangle$ which Ginzburg [11] calculates in connection with linear absorption for oblique incidence. Using his expression (85) becomes (replacing $\partial/\partial x$ by x_0^{-1} since $\langle E_x E_y \rangle$ is significant only over the resonant region)

$$\frac{\partial \langle B_z \rangle}{\partial t} = \left(\frac{\omega}{vx_0}\right)^2 \frac{1}{en_0} \frac{\varepsilon_0 E_0^2}{2} \sin \theta_0 \cos \theta_0 \exp\left[-\tfrac{4}{5}k_0 L \sin^3 \theta_0\right] \qquad (86)$$

which maximizes under the same condition (75) as a function of the angle of incidence θ_0 as the linear absorption.

The electromagnetic term is an order of magnitude larger than the thermoelectric term.

The authors estimate the maximum magnetic field occurring at oblique incidence as follows: Electron expulsion from the resonant region will eventually generate an opposing current which will limit the growth. The maximum field is then given by

$$\langle B_z \rangle \sim \frac{1}{I_x x_0} \frac{\varepsilon_0 \langle E_x E_z \rangle}{2} \qquad (87)$$

where I is estimated from computer simulations to be given by

$$I_x \sim 10^{-2} n_c e. \qquad (88)$$

For neodynium laser light of intensity $2 \cdot 2 . 10^{16}$ W cm^{-2} and $T = 5$ kV, $\langle \partial B_z / \partial t \rangle = 6$ MG (ps)$^{-1}$ and $\langle B_z \rangle = 3$ MG. Such fields would cause a reduction of the rate of heat conduction by $(\omega_c \tau_e)^2 = 700$ which would be a serious obstacle to compression. Normal incidence, i.e. spherical symmetry should greatly reduce the magnetic field.

The existence of such fields was confirmed by microwave experiments. Raven and Rumsby [22] have made measurements in the afterglow of a strong gas discharge. They used microwaves with wavelength 1·8 cm and they were able to change the polarization and the angle of incidence on a linear density gradient. For the parameters of their experiment theory predicts a growth rate of 30 G μs^{-1} which is in reasonable agreement with the measured value of 18 G μs^{-1}. Magnetic field measurements were also reported by DiVergilio et al. [23].

Further theoretical work may be found in [18–20]. In laser experiments, very high magnetic fields were reported by Stamper et al. [16] and by Raven et al. [17] in plane target experiments. No magnetic fields have been found as yet for spherical targets under conditions of uniform illumination. Further literature on the subject may be found in [21, 22]. The experimental facts do not, as yet, fall into a clear pattern and for this reason a detailed account is not attempted.

References

1. Dawson, J. and Oberman, C. (1962). *Phys. Fluids*, **5**, 517.
   Dawson, J. and Oberman, C. (1963). *Phys. Fluids*, **6**, 394.
2. Johnstone, T. W. and Dawson, J. M. (1973). *Phys. Fluids*, **16**, 722.
3. Kaw, P. and Dawson, J. M. (1969). *Phys. Fluids*, **12**, 2586–2591.
4. de Groot, J. S. and Katz, J. I. (1973). *Phys. Fluids*, **16**, 401–407.
5. Hughes, T. P. (1975). "Plasmas and Laser Light". Adam Hilger, London.
6. Kruer, W. L. and Dawson, J. M. (1972). *Phys. Fluids*, **15**, 446–453.
7. Budden, K. G. (1961). "Radio Waves in the Ionosphere". Cambridge University Press, Cambridge.
8. White, R. B. and Chen, F. F. (1974). *Plasma Phys.* **16**, 565–587.
9. Denisov, N. G. (1957). *Soviet Phys JETP*, **4**, 544.
10. Freidberg, J. P., Mitchell, R. W., Morse, R. L. and Rudsinski, L. I. (1972). *Phys Rev. Lett.* **27**, 795.
11. Ginzburg, V. L. (1970). "The Propagation of Electromagnetic Waves in Plasmas", 2nd Edn. Pergamon Press, Oxford.
12. Morales, G. and Lee, Y. C. (1974). *Phys Rev. Lett.* **33**, 1016.
    Morales, G. and Lee, Y. C. (1976) *Phys. Fluids*, **19**, 690.
13. Kim, H. C., Wong, A. Y. and Stenzel, R. L. (1974). *Phys. Rev. Lett.* **33**, 886.
14. Wong, A. Y. (1977). "Laser Interaction and Relate Phenomena" (H. J. Schwartz and H. Hora, eds), Vol. IV B. Plenum Press, New York.
15. Thompson, J. J., Max, C. E. and Estabrook, K. (1975). *Phys Rev. Lett.* **35**, 663.
16. Stamper, J. A., Papadopoulos, K., Sudan, R. N., McLean, E. A. and Dawson, J. M. (1971). *Phys Rev. Lett.* **26**, 1012.
17. Raven, A., Willi, O. and Rumsby, P. T. (1978). *Phys. Rev. Lett.*, **41**, 554.
18. Colombant, D. G. and Winsor, N. K. (1975). *Phys. Rev. Lett.*, **38**, 697.
19. Tidman, D. A. (1974). *Phys. Rev. Lett.*, **32**, 1179.
20. Craxton, R. S. and Haines, M. G. (1975). *Phys. Rev. Lett.*, **35**, 1336.
21. Max, C. E., Manheimer, W. M. and Thomas, J. S. (1978). *Phys. Fluids*, **21**, 128.
22. Raven, A. and Rumsby, P. T. (1977). *Phys. Lett.*, **60** A, 42.
23. Di Vergilio, W. F., Hong, A. Y., Kim, A. C. and Lee, Y. C. (1977). *Phys. Rev. Lett.*, **38**, 541.
24. Williamson, H. (ed.) (1976). "Plasma Physics", Plenum Press, New York.
25. Thornhill, S. G. and ter Haar, D. (1978). *Physics Reports* **43**, 43.

8

TRANSMISSION OF LASER POWER
TO THE TARGET

In Chapter 7, we showed that the absorption region is located in a plasma density range near the critical density. For laser wavelengths of approximately 1 μm this is a density of $\approx 10^{21}$ per cm^3 whereas solid state density for hydrogen is $\approx 10^{23}$ per cm^3. The important question which arises concerns the efficiency of the use of laser power for the compression of the target.

First, we cast a brief look at the sequence of events. Laser power impinging on the target surface is initially absorbed by the production of photoelectrons. This happens by multiphoton absorption and the electrons in turn ionize and heat the target. The target material evaporates, the vapour is ionized by collisions with laser accelerated electrons and a plasma layer is formed. The density of this plasma near the target surface is much above critical density, and the absorption occurs at progressively larger distances from the surface. Hora [20] has considered laser intensities so high that the electric field may cause non-collisional heating and outward acceleration of plasma in an explosive fashion. It is not clear whether this process is realizable. Thus setting this possibility aside, the plasma between the absorption region and the solid is heated by conduction. The physics of heat conduction will be discussed below and it will be seen that the classical theory is not always applicable.

The plasma layer expands, mostly in an outward direction. Meanwhile the heat is conducted into the solid and leads to further evaporation. This process is called ablation. The momentum of the ablated plasma is balanced by momentum transfer to the solid. If the radiation arrives on a freely suspended plane target, it is accelerated and this process has been studied by Burgess et al. [4], experimentally. If the irradiation is spherically uniform, compression of the target results and has been observed by many authors, as will be reported in Chapter 13.

The whole process is at present far from being understood and we shall, at first, present simple dimensional arguments and rather primitive theories leading to estimates of the power transferred to the solid in the ablation process. It will be seen that it is only a small fraction of the incident power.

The momentum transferred by the action of the ablating gas pressure and the effect of the ponderomotive force launches a shock wave in the solid which after a short time overtakes the front of the heat wave.

Adopting a rather crude approximation to Fourier's law of heat conduction the depth l_H to which energy absorbed at the rate ϕ W cm^{-2} has penetrated after a time t is given by

$$l_H = (\kappa t)^{1/2} \tag{1}$$

where κ is the heat conductivity. Using a law $\kappa = bT^n (n > 0)$ and energy balance

$$\phi t = \tfrac{3}{2} N k T_e l_H, \qquad l_H \propto t/T_e = t \left(\frac{\kappa}{b} \right)^{-1/n} \tag{2}$$

$$l_H \propto t^\alpha, \qquad \alpha = \frac{n+1}{n+2} < 1. \tag{3}$$

Now let us estimate the penetration length l_E of the shock wave moving with sound velocity C_s

$$l_E \approx C_s t \propto T_e^{1/2} t \propto t^\beta, \qquad \beta = \frac{2n+5}{2n+4} > 1. \tag{4}$$

This is illustrated in Fig. 1 showing the shock wave reaching the heat wave in a finite time t_0 at a distance l_0.

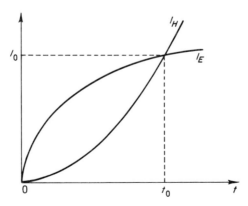

Fig. 1. Penetration of heat and shock wave into plane target. Shock overtakes temperature wave at depth l_0 at time t.

Following Caruso and Gratton [1] we now determine l_0 and t_0. For a sufficiently high laser intensity ionization and dissociation are unimportant and we take transport coefficients calculated for a plasma:

$$\kappa = bT^{5/2} \tag{5}$$

on the assumption that the classical theory of heat conduction discussed below is valid.

Precise values for the constant b in terms of thermoelectric coefficient are given below (27, 28).

From (1) to (5)

$$\phi t = \tfrac{3}{2}Nk T_e l_H = \tfrac{3}{2}Nk \left(\frac{l_H^2}{tb}\right)^{2/5} l_H, \qquad l_H^{9/5} = \frac{\phi t^{7/5} b^{2/5}}{\tfrac{3}{2}(Nk)}$$

and it follows that

$$l_H \equiv b^{2/9}(\tfrac{3}{2}Nk)^{-7/9}\phi^{5/9}t^{7/9}$$

$$T_e \equiv b^{-2/9}(\tfrac{3}{2}Nk)^{-2/9}\phi^{4/9}t^{2/9}$$

and from (4) for ions with mass M

$$l_E = \left(\frac{2kT_e}{M}\right)^{1/2} t = \phi^{2/9}t^{10/9}$$

equating $l_0 = l_H(t_0) = l_E(t_0)$ it is found that

$$t_0 = \left(\frac{M}{2k}\right)^{3/2} b(\tfrac{3}{2}kN)^{-2}\,\phi \tag{6}$$

$$l_0 = \left(\frac{M}{2k}\right)^{3/2} b(\tfrac{3}{2}kN)^{-7/3}\phi^{4/3} \tag{7}$$

and it is seen that both l_0 and t_0 increase with ϕ.

Numerically with $b = 2.10^{-13}$, $N = 5.10^{22}\,\text{cm}^{-3}$ for hydrogen

$\phi(\text{W cm}^{-2})$	10^{12}	10^{14}	10^{16}
$t_0(\text{s})$	10^{-13}	3.10^{-12}	3.10^{-10}

The time t_0 is an interesting quantity. If the laser pulse is short compared to t_0 the heated matter has no time to move and the energy diffuses into the target by heat conduction. If, on the other hand, the laser pulse lasts for a time large compared to t_0, the expansion process happens while the light is being absorbed.

When the laser pulse length $T \gg t_0$ we must distinguish three phases as illustrated by Fig. 2; phase 0 of mass density ρ_0 of the undisturbed solid; phase 1 of density ρ_1 which we shall call the dense phase of opaque ionized matter heated by the shock wave and moving into phase 0 with shock speed v_s while the mean particle velocity is v_1; phase 2 composed of plasma sublimated from the boundary of phase 1. For this phase we can expect a kind of self-regulating regime. If it becomes too opaque the laser light cannot penetrate and cannot produce it, if it becomes too transparent more plasma will be sublimated. The boundary between phase 1 and phase 2, the ablation front moves with speed v_F.

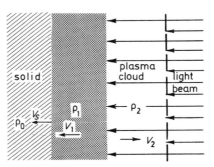

Fig. 2. Regions of ablating plasma (schematic).

We can write down momentum and energy balance neglecting the pressure in phase 0 and denoting the pressure acting on the ablating surface tending to compress the dense phase by $P = \rho_2 v_2^2$:

$$-\rho_0 v_s = \rho_1(v_1 - v_s), \qquad \rho_0 v_s^2 = \rho_1(v_1 - v_s)^2 + P. \tag{8}$$

This gives

$$v_1^2 = \left(1 - \frac{\rho_0}{\rho_1}\right)\frac{P}{\rho_0}, \qquad v_s^2 = \frac{\rho_1}{\rho_1 - \rho_0}\frac{P}{\rho_0}. \tag{9}$$

Hence $v_1^2 \leqslant P/\rho_0$ and this gives an upper limit to the flux transferred to the dense phase

$$P|v_1| \leqslant P\left(\frac{P}{\rho_0}\right)^{1/2} \approx \rho_2 v_2^2 \left(\frac{\rho_2 v_2^2}{\rho_0}\right)^{1/2} = \rho_2 v_2^3 \left(\frac{\rho_2}{\rho_0}\right)^{1/2} \approx \phi\left(\frac{\rho_2}{\rho_0}\right)^{1/2} \tag{10}$$

($\rho_2 v_2^2 \cdot v_2$ represents energy moving with velocity v_2).

But $\rho_2/\rho_0 \ll 1$ and so only a small fraction of the energy is transferred to phase 1 to be available to compress the target.

Caruso and Gratton [1] derived scaling laws on the assumption that phase 2 is transparent to the light.

The absorption law is given by formula (43) of Chapter 7. We can write

$$\left|\frac{d\phi}{dx}\right| = a\rho_2^2 v_2^{-3}\phi \tag{11}$$

where a is a constant. We are interested in the mean expansion velocity v_2 and the density ρ_2 neglecting any effect on the nature of the plasma produced and we assume plane geometry. There are only two independent dimensional quantities a and ϕ which we shall use together with the time t. The dimensions of a and ϕ are

$$[a] = l^8/t^3m^2, \qquad [\phi] = m/t^3. \tag{12}$$

The ratio of the length l filled by the plasma and the penetration depth s will be a constant and in order to form a velocity of dimension l/t we must have

$$t^x a^\beta \phi^\gamma = l/t$$

$$8\beta = 1, \qquad \alpha - 2\beta + 3\gamma = -1, \qquad -2\beta + \gamma = 0$$

which yields $\alpha = \frac{1}{8}, \beta = \frac{1}{8}, \gamma = \frac{1}{4}$ and

$$v_2 = K_v(at)^{1/8}\,\phi^{1/4} \tag{13}$$

similarly

$$\rho_2 = K_\rho(at)^{-3/8}\phi^{1/4} \tag{14}$$

where K_v and K_ρ are constants. The evaporated mass per unit area,

$$\mu = \int^t \rho_2 v_2\,dt = K_\mu a^{-1/4}t^{3/4}\phi^{1/2} \tag{15}$$

and the pressure P

$$P = \rho_2 v_2^2 = K_p(at)^{-1/8}\phi^{3/4} \tag{16}$$

can be related to a, ϕ and t. The constants K_v and K_p turn out to be near to unity. For plane targets and $\phi < 10^{13}\ \text{W cm}^{-2}$, the above scaling laws have been experimentally verified [4]. Other scaling laws applicable under different conditions have been given by Caruso [2] who also presented a general review of scaling in connection with model experiments for laser driven hydrodynamics [3].

The problem of the power transfer to the pellet is an important one. Since the laser radiation cannot penetrate much beyond the critical layer where its frequency equals the plasma frequency, we must distinguish between the outer region where the electrons are directly heated to a temperature T_h which is called the corona and the inner region at temperature T_{co} the "core" where they are not. Some of the laser light will be reflected from the corona region and the heated electrons from the corona will be cooled by the core. This is schematically illustrated in Fig. 3.

If the cooling is not effective, the corona will be heated and expand,

F

escaping mostly in an outward direction before the energy is transferred to
the core, and the corona and core will be decoupled. Kidder and Zink [5]
drew attention to the effect of wavelength; for the longer wavelength (10 μm)
of a CO_2 laser, the location of the critical wavelength region is so much further
away from the core that decoupling may result and they confirm this conclusion with a very simplified analysis. A proper theoretical treatment of the

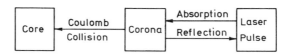

Fig. 3. Schematic illustration of corona region (absorbing laser light) cooled by ablating core
material.

energy transfer to the target does not as yet exist. It would have to treat
dynamically, the movement of the ablation front, and the energy and momentum balance between incoming laser light, reflected light and ions and electrons. Many numerical simulation studies have been carried out which refer to
particular situations. We shall content ourselves with an outline of the
physical processes involved.

The following important questions arise:

(i) What is the dominant absorption mechanism?

An absorption coefficient calculated on the basis of the inverse *bremsstrahlung* is calculated in Chapter 7. This calculation is valid in a uniform medium
but the expanding plasma cloud is non-uniform and additional absorption
processes occur: resonant absorption, different for normal, and oblique
incidence as discussed in Chapter 7; and absorption due to parametric
instabilities converting laser energy into plasma wave engery, which in turn
is transformed into more or less random particle energy. For higher laser
power density the ponderomotive force plays an increasingly important part
and modifies the density profile leading to cavities and particle acceleration
as explained in Chapters 6, 7 and 13. Particle acceleration may also occur in
resonant absorption process through "wave breaking". Albritton and Koch
[7] pointed out that in the case of oblique incidence, particles in the resonance
region oscillate in the direction of the density gradient with an amplitude
increasing with time. The space charge force acts as a restoring force as long
as the electrons stay on the same side of the resonance position. With
increasing amplitude, electron orbits cross, and the electrons get into a field
region which continues to accelerate them and drives them out of the resonance region. The phenomenon is called wave breaking because steep fronts
of the kind discussed in the following chapter are formed when particle orbits
cross.

(ii) What is the fraction of the incident laser energy leading to an ion flux streaming in the outward direction?

In many experiments the energy carried away by the ions has been shown to be a large fraction of the incident laser energy. What is the velocity spectrum of the ions and what is the explanation of the observed spectra?

(iii) As will be reported in Chapter 13, fast electron groups are observed with energies of an order of magnitude larger than the electron thermal energy of corona electrons. Do they owe their origin to parametric decay processes, to resonant absorption processes or to acceleration by the ponderomotive force?

(iv) What is the nature of the energy transport from the absorption region to the core? First there is the classical heat conductivity treated by Spitzer and Braginskii [6], valid in a uniform medium. In the expanding plasma the scale length of the inhomogeneity may be short compared to the electron mean free path and this theory is no longer valid. The classical formula would lead to heat conduction far in excess of the maximum transport which can be effected by free streaming electrons. One speaks of flux limitation.

The energy transport may also be inhibited by magnetic fields. In Chapter 7, the generation of high magnetic fields is discussed and it is shown that it occurs predominantly for oblique incidence. In the case of compression of spherical targets it may be expected that magnetic fields are not produced under conditions of uniform illumination but are likely to appear when it is non-uniform.

According to Braginskii [6], in the presence of a magnetic field, the heat conductivity in a direction perpendicular to the field is decreased by a factor $\omega_{ce}\tau_e$, $\omega_{ci}\tau_i$, respectively where ω_{ce}, ω_{ci} are the cyclotron frequencies. This has the effect of inhibiting the heat conduction which would tend to smooth out the non-uniformity. This simple result applies for $\omega\tau \gg 1$ (the τ's are the times $1/\nu$ between collisions).

Absorption, energy conversion and transport in the corona determine the fraction of the incident energy available for vaporizing the solid target and the momentum transfer to the core due to the ablation process. The questions posed are the subject of active research and cannot be answered definitely at the present time. In the light of the comments which follow, it will become clear why a comprehensive theory cannot be offered.

The first comment concerns thermal conductivity; Harrington [16] has shown that the relation

$$q = -\kappa\nabla T \tag{17}$$

between the heat flux q per unit of time and the temperature T gradient can be theoretically deduced for a gas when several conditions are met.

(i) The scale length $L = (1/T)\nabla T$ must be long compared with the mean free path, i.e.

$$L_1 \gg \lambda \tag{18}$$

and a similar condition must be satisfied by the concentration.

(ii) the scale length $L_2 = (1/\nabla T)\nabla^2 T$ must also be large compared to λ:

$$L_2 \gg \lambda. \tag{19}$$

When these conditions are met, the thermal conductivity of a gas becomes

$$\kappa_s = \frac{4N_s k^2 T_s}{\pi m_s}\tau_s \tag{20}$$

where

$$\bar{u}_s = \lambda_s \tau_s = \left(\frac{8kT_s}{\pi m_s}\right)^{1/2} \tag{21}$$

is the mean molecular velocity. The thermal energy $U = c_v T$ satisfies the conservation equation

$$\partial U/\partial t = -\nabla q$$

and the resulting equation

$$c_v \frac{\partial T}{\partial t} = \nabla \cdot (\kappa \nabla T) \tag{22}$$

implies the assumption that

$$|\nabla^2 T| \gg \lambda |\nabla(\nabla^2 T)| \tag{23}$$

since, in effect $\kappa \nabla T$ is expanded in a Taylor series.

Transport coefficients in a plasma are treated comprehensively by Braginskii [6] by means of Boltzmann's equation which shows that the conductivities for ions κ_i and the conductivity κ_e for electrons are given by

$$\kappa_e = \frac{N_e k^2 T_e \tau}{m_e}, \qquad \tau_e = \frac{\bar{u}_e}{\lambda_e} \tag{24a}$$

$$\kappa_i = \frac{N_i k^2 T_i \tau_i}{m_i}, \qquad \tau_i = \frac{\bar{u}_i}{\lambda_i} \tag{24b}$$

$$\tau_e = \tfrac{3}{4}(m_e/2\pi)^{1/2} (4\pi\varepsilon_0)^2 (kT_e)^{3/2}/(Z_e e^4 N_e \ln \Lambda) \tag{25}$$

$$\tau_i = \tfrac{3}{4}\left(\frac{m_i}{\pi}\right)^{1/2} (4\pi\varepsilon_0)^2 (kT_i)^{3/2}/(Z^4 e^4 N_i \ln \Lambda). \tag{26}$$

Putting $N_e = ZN_i$ and inserting the constants

$$\kappa_e = 1\cdot8 . 10^{-10}(T_e^{5/2}/(Z \ln \Lambda)) \, \mathrm{Js^{-1} \, K^{-1} \, m^{-1}} \tag{27}$$

$$\kappa_i = 7\cdot5 . 10^{-12}(T_i^{5/2}/(M^{1/2}Z^4 \ln \Lambda)) \, \mathrm{J \, s^{-1} \, K^{-1} \, m^{-1}} \tag{28}$$

here $\ln \Lambda$ is the coulomb logarithm, the temperatures T_e and T_i of electrons and ions are given in K, and M is the ionic mass in atomic mass units. Salzmann [8] showed that the assumption equivalent to (23) in this case is

$$|\nabla^2 T_e^{7/2}| \gg \lambda_e |\nabla . (\nabla^2 T_e^{7/2})|. \tag{29}$$

(Further formulae valid in the case of inhomogeneous temperature distribution have been given by Braginskii [6]). This follows as before from series expansions since (for constant Λ).

$$K_e \nabla T_e \propto \nabla(T_e^{7/2}). \tag{30}$$

In a plasma, the thermoelectric equations

$$\mathbf{j} = \sigma\mathbf{E} + \alpha\nabla\mathbf{T} \qquad \mathbf{q} = -\beta\mathbf{E} - \kappa\nabla\mathbf{T} \tag{31}$$

must also be taken into account. Here $\mathbf{j}$ is the electric current density and α and β are the coefficients for the Seebeck and Peltier effect (see [4] of Chapter 2) and the effective thermal conductivity in a steady state of flow is given by κ. This has been incorporated into (24a) and (24b).

Bickerton [9] has pointed out that a valid theory of thermal conductivity under the conditions existing in laser fusion experiments, where one or more of the above conditions are not satisfied, does not yet exist. It is clear that the thermal flux cannot exceed that carried by all the electrons moving with their thermal velocity down the temperature gradient given by

$$q_{\max} = N_e(3kT_e/m)^{1/2} \tfrac{3}{2}kT_e. \tag{32}$$

From (17, 24, 20, 21) (with $\nabla T_e \approx T_e/L_e$) it would follow that for small gradients

$$q_e = 0\cdot77 \frac{\lambda_e}{L_e} q_{\max} \tag{33}$$

for $L_e < \lambda_e$ the theory breaks down, just when it would predict fluxes greater than that given by (32). It must be assumed that the flux is limited to a fraction β of $q_{\max}$, i.e. that

$$q_e = \beta q_{\max}. \tag{34}$$

When such a flux limitation is postulated it remains to determine β. Various estimates have appeared in the literature. A careful discussion of experiments by Haas *et al.* [10] concludes that there is strong evidence of transport inhibition and that β must lie between 0·01 and 1.

Concerning (ii) we again turn to the experimental evidence. Here we must distinguish between experiments with foils and perpendicular laser incidence and spherical compression experiments. The conditions are simpler in the former and typical ion spectra from an experiment with polythene foils are given in Chapter 13, Fig. 9. Trying to interpret such spectra theoretically it must be said at once that they do not reveal directly what happens when ions are emitted near the ablation layer; further acceleration processes intervene to produce the observed ion energies.

Consider the expanding plasma leaving the ablation front. It is electrically neutral to start with, but the electrons with their higher speed tend to escape creating a charge unbalance which gives rise to a field tending to decelerate the electrons and accelerate the ions. As the plasma front expands the ions are accelerated by this field. Leaving aside the absorption and electron production by the incident laser light, a model of isothermal expansion is treated by Crow *et al.* [13]. In plane geometry they solve simultaneously continuity, ion momentum equation and Poisson's equation

$$\frac{\partial v}{\partial t} + v\frac{\partial v}{\partial x} = \frac{e}{m_i} E, \tag{35}$$

$$\frac{\partial n_i}{\partial t} + \frac{\partial}{\partial x}(n_i v) = 0 \tag{36}$$

$$\varepsilon_0 \frac{\partial^2 V}{\partial x^2} = e(n_e - n_i) \tag{37}$$

where v is the ion velocity. They make the assumption that the electrons achieve equilibrium in a time small compared to the expansion process and assume a Boltzmann distribution

$$n_e = n_0 \exp(eV/kT_e) \tag{38}$$

for the electrons in the potential $V(x)$. This assumption closes the set of equations and allows numerical computation of the time change of electron density. Starting with a rectangular ion distribution and an electron distribution illustrated in Fig. 4 the computation yields a moving front, illustrated in Fig. 5 leading, by Poisson's equation, to a potential and field distribution illustrated by Fig. 6. The field continuously accelerates the ions and the resulting distribution is shown in Fig. 7 at various times (measured in terms of the ion plasma frequency). Here $n_e = n_i$ except at the positions where the

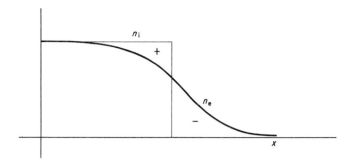

Fig. 4. Plasma before expansion. (From [13], with permission.)

ions form a front. Unfortunately, this model predicts indefinite accelerations of the ions and it is clear that it must break down at some stage. The equation which fails is the Boltzmann equilibrium equation (38) which certainly does not hold when the ion velocity approaches the electron thermal velocity. The authors suggest that it may break down when $v = \bar{u}_e/4$. The kinetic energy of the ions then becomes

$$\tfrac{1}{2}m_i v^2 = (m_i/4\pi m_e)\,\mathrm{k}T_e. \tag{39}$$

For an electron temperature of 100 eV, the ion energy would be 15 keV which is of the order of magnitude of the observed energies.

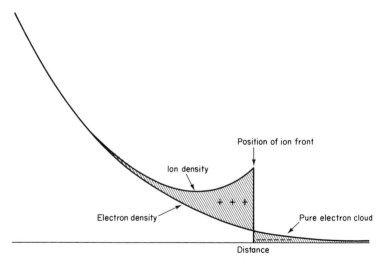

Fig. 5. Variation of electron and ion densities at the front. (From [13], with permission.)

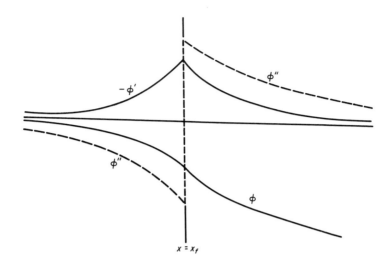

Fig. 6. Variation of potential ϕ, first derivative ϕ' and second derivative ϕ'' through the ion front. (From [13], with permission.)

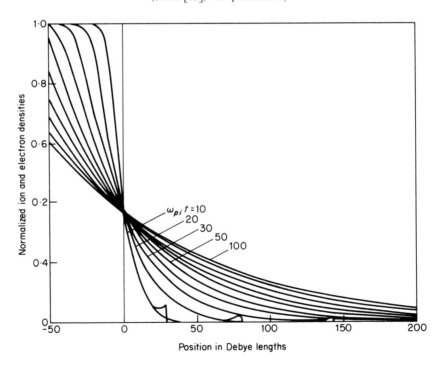

Fig. 7. Ion and electron densities against position at several times. (From [13], with permission.)

The equilibrium assumption (38) is questionable, so kinetic equations should be employed instead. If we keep the concept of the electric field near the front, then slow electrons will be reflected from the resulting potential ramp and fast electrons will carry on. This might of course lead to some turbulence in the counterstreaming region. The electron distribution of the electron precursor will be non-Maxwellian. Such an analysis has not yet been carried out, neither has the problem, as formulated above, been worked out for the case of spherical expansion. Experiments with plane targets yield spectra compatible with the theory. Crow et al. [13] show that their theory provides a modification of a similarity solution, not involving space charge, obtained by Gurevich et al. [12]. This free streaming solution according to which

$$n = n_0 \exp[-(x + ct)/ct], \qquad v = (x + ct)/t \qquad (40)$$

$$V = -(kT_e/e)(x + ct)/ct, \qquad E = (kT_e/ect) \qquad (41)$$

where $c = (kT_e/m_i)^{1/2}$, E the electric field, V the potential obtained by these authors for $x > -ct$ may well serve as an interpretation of the observed proton distribution given by Lindman ([4] of Chapter 13) shown in Fig. 9 of Chapter 13. The peak at higher speeds may be due to the modification caused by the space charge, and predicted by the results in [12].

All this is further modified when the absorption of laser light is considered. Moreover, at high incident power the density profile modification due the ponderomotive force in the absorption region becomes important and one must look for further clues to elucidate the situation. Some of these might be provided by computer simulations, but the conclusions from the existing literature are too complex to be summarized here. Other clues may be provided by careful observation. Most experiments reveal a two-temperature distribution of electron velocities. Allen and Wickens [17] have studied the effect of the hot and cold electrons in this two-temperature distribution on the ablating plasma with a self-similar analysis of the fluid equations in plane geometry (equations (35) and (36)). They do not take into account Poisson's equation (37) but assume quasineutrality;

$$Zn_i = n_h + n_c. \qquad (42)$$

where n_h is the hot electron number density and n_c the cold electron number density. With

$$n_h = n_{h0} \exp(eV/kT_h), \qquad n_c = n_{c0} \exp(eV/kT_c) \qquad (43)$$

as approximations to the experimentally observed electron distributions (cf. Estabrook and Kruer [19]) they find that

$$v_i - x/t = S(V) \qquad (44)$$

and

$$dv/dV = -Ze/[m_i S(V)] \tag{45}$$

where

$$S(V) = \{Z(n_h + n_c)/(m_i[n_h/kT_h + n_c/kT_c])\}^{1/2} \tag{46}$$

is the ion sound speed. (Note that in (40, 41) the sound speed is constant.) Equation (45) can be integrated to give the ion velocity as a function of the potential. Inverting this and using (43) and (42) the velocity distribution of the emitted ions is found as a function of the ion velocity.

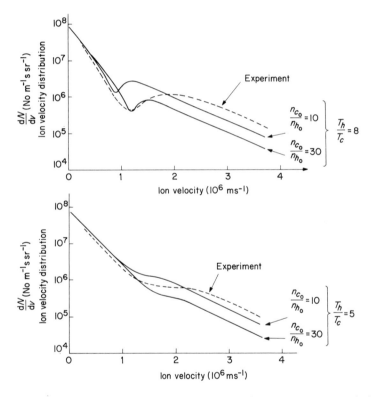

Fig. 8. Ion velocity distribution for various ratios T_h/T_c. Experimental points are marked which fit the theoretical curve.

Figure 8 shows plots of ion velocity distribution for two values of the temperature ratio T_h/T_c. The theory predicts that the slopes of the lower and upper asymptotes are proportional to the inverse square root of the hot and cold electron temperatures respectively, and that from the size and position of the dip in the curve, the ratios T_h/T_c and n_h/n_c can be estimated. The values for the temperature ratios found in this manner are consistent with X-ray data, such as reported in Chapter 13 (see e.g. Fig. 6 of Chapter 13) and the density ratios are consistent with order of magnitude estimates. An interesting feature is that for $T_h/T_c > 10$ the solution can become multi-valued. Bezzerides et al. [21] have fitted an expansion shock to avoid this. A comparison with experiment is reported in [18].

Turning to the experimental results concerning the production of hot electrons, we refer to a compilation of data from many experiments carried out in several laboratories illustrated by Fig. 7 of Chapter 13. In this figure the temperature of the hot electrons is plotted against the product of the laser flux and the square of the wavelength. The results lie on a universal curve showing a scaling

$$T_H \propto (\phi\lambda^2)^\delta \tag{47}$$

where V_H is the hot electron temperature and δ depends on the $\phi\lambda^2$ range.

A tentative interpretation of these data may be given as follows. Assuming that a fraction α_0 of the incoming light energy is absorbed and that the flux-limited transport regime applies than

$$\alpha_0\phi = \beta n_H v_H m_e(v_H^2/2) \tag{48}$$

where v_H is the velocity of the hot electrons, n_H their density and β the flux limitation factor introduced earlier. With α_0 given by equation (43) of Chapter 7, T_H then scales as

$$T_H \propto (\phi\lambda^2)^{2/3} \tag{49}$$

giving nearly the value for δ measured for $10^{13} < \phi\lambda^2 < 10^{15}$ W μm² cm^{-1}.

For higher flux densities pressure balance

$$\varepsilon_0 E^2/2 = nT_c \quad (T_c \text{ in keV}) \tag{50}$$

may be reached. Here the appropriate temperature is that of the cold plasma electrons, T_c, rather than the temperature T_h of the hot ones which constitute only a fraction of the electron population. Data concerning the ambient temperature of the plasma will be found in Chapter 13. The density gradient is steepened and the laser light penetrates to a density region $n = \varepsilon_0 E^2/2T_c$ which for a laser power density of 10^{15} W cm^{-2} is 10^{21} for both CO_2 and Nd lasers. The electric laser field drives the electrons in the high gradient region which has a length scale $L_s = c/\omega_p = 0\cdot16$ μm where ω_p is the corresponding plasma frequency. This leads to a hot electron temperature

$$V_H = EL_s \propto ET_c^{1/2}/\phi^{1/2} \propto T_c^{1/2} \tag{51}$$

so that V_H scales as $T_c^{1/2}$ independent of laser power and wavelength! This is not observed but the scaling does become weaker and computer simulations by Forslund *et al.* [14] lead to a value of $\delta = 0.33$ under similar assumptions.

This kind of explanation applies only to the plane target experiments quoted. For spherical targets and uniform illumination, the incidence is normal and the radial component of E needed for resonance acceleration in the radial direction should not arise. If the weak absorption due to inverse *bremmstrahlung* is to be exceeded, nonlinear wave conversion must be invoked. In a high-density gradient, the threshold of Brillouin scattering is raised, but back-scattering does not become excessive unless a uniform density plasma is created by a weaker pre-pulse or by long pulses at high intensity which spread the plasma. It will be seen in Chapter 13 that Brillouin scattering processes are almost certainly involved but it is not easy to distinguish them from resonance absorption. A radial component of the electric field as needed for resonance absorption may arise if the plasma surface is rippled. Attention was drawn in Chapter 6 to simulations by Estabrook ([10] of Chapter 6) which show a complicated structure, and evidence for rippling from holographic studies will be reported in Chapter 13. The scaling law (4), deserves a further comment. $\phi\lambda^2$ is proportional to the energy of the electrons acquired in the field of the laser. It is therefore natural to conclude, that the change in the rate of rise of the hot electron energy is connected with steepening of the density profile, which, according to our discussion of plasma-field equilibria, is governed by the ratio of the field energy to the thermal energy of the plasma. There have been several attempts to formulate a theory of profile modification on these lines, e.g. by Virmont *et al.* [15]. We shall see in Chapter 13 that there is evidence for profile steepening but it would seem that the problems of rippling and steepening cannot be separated.

Two-plasmon decay which occurs at a position where the density is one quarter of the critical density must also be considered. Emission of light at frequencies of 2ω and $\frac{3}{2}\omega$ is observed and this may be the result of nonlinear combination of the decay product, the Langmuir turbulence with the laser frequency. Harmonics of the incident light may also be due to the nonlinear conversion in a steep density gradient.

As has been mentioned before, the hot electrons emitted in a direction towards the target may heat it to such an extent that the work needed for compression is very much increased, so that very large laser power is needed. For ablatively driven compression the power density must be kept low enough, so that the resulting hot electrons do not penetrate the target. Thus one must use target diameters larger than those used for experiments where the "exploding pusher type" of compression is attempted. In the latter case, thin small targets are used. The hot electrons penetrate the shell and evaporate it rapidly. Many such experiments use glass balloons filled with DT mixture.

In this case, it is not hard to achieve symmetric implosion because the hot electrons heat the shell uniformly. On the other hand, a useful fusion yield can be achieved only with great energy economy and ablatively driven implosion without preheating is required. It is thought that the necessary symmetry can only be achieved by the convergence of a multiplicity of beams. As it is clearly necessary to realize ablative compression we now formulate some of the conditions which must be fulfilled in addition to uniform illumination. The range R of the hot electrons must, as we have seen be smaller than the thickness of the shell, Δr_0. Very roughly $R = T_H^2/\rho$ μm where T_H is expressed in keV. We set

$$\Delta r_0 = pT_H^2/\rho = p\phi^{2\delta}/\rho \tag{52}$$

where p is a number larger than unity and the hot electron energy is given by

$$T_H \propto \phi^\delta, \qquad \mathbf{q} \approx 0.3. \tag{53}$$

The force f which accelerates the ablating ions is proportional to T_H, i.e. the suprathermal pressure, and thus acceleration scales as

$$a \propto T_H/(\rho\Delta r_0) \tag{54}$$

while the velocity of the ablating ions is proportional to $(r_0 f)^{1/2}$. The compression takes a time t which varies according to

$$t \propto (r_0/f)^{1/2}. \tag{55}$$

Taking into account (52) and (53), a now scales as

$$a \propto \frac{T_H}{\rho\Delta r_0} = \frac{1}{p\phi^\delta} \tag{56}$$

and time as

$$t \propto (r_0 p\phi^\delta)^{1/2} \tag{57}$$

while the energy density ε varies as $v^2 \propto r_0/(p\phi^\delta)$.

We must postulate that not all of the shell ablates away, i.e. that

$$Cm_i n_c t/Z < \rho\Delta r_0, \qquad \text{i.e.} \quad Ct < \Delta r_0 \frac{\rho Z}{n_c m_i} \tag{58}$$

where C is the ablation velocity which, if the process is caused by the hot electrons may be given by $\frac{1}{2}m_i C^2 n_c = T_h n_H$ or

$$C = (T_H/m_i)^{1/2}(n_H/n_c)^{1/2} \tag{59}$$

where m_i is the mass of the ablating ions. The density n_c is the critical electron density for the laser wavelength, $Zn_i = n_c$, n_H the number of hot electrons.

This leads to

$$\Delta r_0 > (r_0 p)^{1/2} \phi^\delta (m_i \cdot n_H n_c / \rho Z).$$
(60)

and since, at constant laser energy $\phi r_0^2 =$ constant

$$(\Delta r_0)_{min} \propto \phi^{\delta - 1/4} p^{1/2} q, \quad q = (m_i \cdot n_H n_c)^{1/2} / \rho Z$$
(61)

and this depends only weakly on ϕ. On the other hand, we have learned that the growth of the Rayleigh Taylor instability is determined by the growth exponent (cf. Chapters 2 and 10)

$$\gamma t = (r_0 / \Delta r_0)^{1/2}$$
(62)

We have seen that $(\Delta r_0)_{min}$ and t are large for large p and the energy density is low. Thus thick shells and long pulses should produce cool dense implosions provided it is possible to avoid instability. Hora [20] has considered short rise times of high intensity laser pulses (reaching 2.10^{16} W cm^{-2}) in a time less than 10^{-13} s and finds standing waves of high amplitude near the critical density. He assumes that the effective electron temperature is given by

$$T = T_{th} + \frac{e^2 E^2}{m_e \omega^2 K}$$
(63)

where the second term comes from the electron oscillations in the field E of the laser. In the plasma the laser field E is enhanced on account of the refractive index and is given by

$$E^2 = \frac{E_v}{(n_r)^2}$$
(64)

where E_v is the vacuum field. The refractive index in turn depends *inter alia* on the temperature through the collision frequency which is proportional to $T_e^{-3/2}$. Using the enhanced temperature he found large "swelling factors" $|n_r|^{-1}$ up to $n_r^{-1} = 400$. One would then expect ion energies corresponding to the corresponding quasi-potentials. Ions would then flow inward and outward, the group moving towards the centre of the pellet causing compression. In this way, Hora computed a 23% utilization of laser power. It is not clear whether the assumptions made by Hora can be realized in practice.

REFERENCES

1. Caruso, A. and Gratton, R. (1968) *Plasma Phys.* **10**, 867.
2. Caruso, A. (1975) *Plasma Phys.* **18**, 241–242.
3. Caruso, A. (1977). CNEN Frascati, Rome, refs CP 65.
4. Burgess, M. D. J., Motz, H., Rumsby, P. T. and Michaelis, M. (1977). Plasma Physics and Controlled Nuclear Fusion Research, Berchtesgaden, 405–412. Intern. Atom. Energy Agency, Vienna.
5. Kidder, R. E. and Zink, J. W. (1972). **12**, 325.
6. Braginskii, S. I. (1965). *In* "Reviews of Plasma Physics" (M. A. Leontovich, ed.), Vol. I. Consultants Bureau, New York.
7. Albritton, J. and Koch, P. (1975). *Phys. Fluids*, **18**, 1136.
8. Salzmann, H. (1972). *Phys. Lett.* **41** A, 363.
9. Bickerton, R. J. (1973). *Nucl. Fus.* **13**, 457.
10. Haas, R. A., Mead, W. C., Kruer, W. L., Phillion, D. W., Kornblum, H. N., Lindl, J. D., MacQuigg, D. R., Rupert, V. C. and Tirsell, K. G. (1977). *Phys. Fluids*, **20**, 322–338.
11. Allen, J. E. and Andrews, J. G. J. (1970). *Plasma Phys.* **4** 187–194.
12. Gurevich, A. V., Priiskay, L. V. and Pitaevskii, L. P. (1966). *Sov. Phys JETP*, **22**, 449.
13. Crow, J. E., Auer, P. L. and Allen, J. E. (1975). *J. Plasma Phys*, **14**, 65–76.
14. Forslund, D. W., Kindel, J. M. and Lee, K. (1977). *Phys Rev. Lett.*
15 Virmont, J., Pellat, R. and Mora, P. (1977). *Phys. Fluids*,
16. Harrington, R. E. (1967) *J. Appl. Phys.* **38**, 3266.
17. Wickens, L. M. and Allen, J. E. (1979). *J. Plasma Phys* in press.
18. Wickens, L. M., Allen, J. E. and Rumsby, P. T. (1978). *Phys. Rev. Lett.* **41**, 243.
19. Estabrook, K. and Kruer, W. L. (1978). *Phys Rev. Lett.* **40**, 42–45.
20. Hora, H. (1977). *In* "Laser Interaction and Related Phenomena" (H. J. Schwarz and H. Hora, eds.), Vol. IV. Plenum Press, New York.
21. Bezzerides, B., Forslund, D. W. and Lindman, E. L. (1978). *Phys. Fluid.* **21**, 2179–85.

9

SHOCK WAVES AND IMPLOSIONS

In this chapter, we present an introduction to nonlinear wave propagation, in so far as it is relevant to our subject. The simple ideas presented at first will, it is hoped, provide an intuitive understanding which will be helpful in dealing with the more complicated phenomena which occur when matter is rapidly compressed by laser-induced ablation pressure, and a succession of waves is launched into the uncompressed solid. We start by following Whitham's treatment [5] closely.

$$\rho_t + c_0 \rho_x = 0, \qquad c_0 = \text{constant} \tag{1}$$

where ρ is written as the dependent variable to fix attention on density waves, although other quantities may show similar wave behaviour. Subscripts indicate partial differentiation with respect to the time t and one space variable x. The solution to (1) is $\rho(x, t) = f(x - c_0 t)$ where $f(x)$ is an arbitrary function which has to be matched to prescribed initial or boundary data. Wave motion governed by this equation thus translates a profile $f(t)$ unchanged from, say $x = x_0$ to another place $x = x_1$ with a delay given by $(x_1 - x_0)/c_0$ so that c_0 is the speed of the wave.

A simple variant of (1) is

$$\rho_t - c_0 \rho_x = 0 \tag{1a}$$

with solution $f(x + c_0 t)$ which travels in the opposite, namely the negative x-direction. The reader may be more familiar with the wave equations of the second order

$$c_0^2 \phi_{xx} - \phi_{tt} = \left(c_0 \frac{\partial}{\partial x} - \frac{\partial}{\partial t} \right) \left(c_0 \frac{\partial}{\partial x} + \frac{\partial}{\partial t} \right) \phi = 0$$

which have both types of solution, the forward and the backward travelling one.

Although (1) is a rather simple equation

$$\rho_t + c(\rho)\,\rho_x = 0 \tag{2}$$

is not, $c(\rho)$ is now a given function of ρ.

Equation (2) may be written as an ordinary rather than a partial differential equation by noticing that the left side is the total derivative $d\rho/dt$ along a curve which has a derivative

$$dx/dt = c(\rho). \tag{3}$$

For any curve in the (x, t) plane we may consider ρ and x to be functions of t and the total derivative of ρ is

$$\frac{d\rho}{dt} = \frac{\partial \rho}{\partial t} + \frac{dx}{dt}\frac{\partial \rho}{\partial x}.$$

It follows that if one only knew a curve C in the (x, t) plane which satisfies (3) then (2) would tell us that $d\rho/dt = 0$ and therefore ρ remains constant along it. Thus C is a straight line with slope $c(\rho)$ and the solution of (2) consists of a family of straight lines with different constant slopes corresponding to different values of ρ. We have chosen the letter C to denote curves of this family because they are called "characteristics".

Since ρ is constant on such a line we could construct them all if we knew a starting value for ρ on each. This is where initial or boundary data come in. Supposing that initial data at $t = 0$ are given by

$$\rho(x, 0) = f(x) \tag{4}$$

we can proceed to construct the lines as follows.

If one of the lines intersects $t = 0$ at $x = \xi$ then $\rho = f(\xi)$. Therefore the line starting at $x = \xi$ has a slope

$$\frac{dx}{dt} = c\{f(\xi)\} \equiv F(\xi) \tag{5}$$

and the equation of the line is

$$x = \xi + F(\xi)\,t. \tag{6}$$

This is illustrated in Fig. 1 where straight lines starting at various points $x = \xi_1, x = \xi_2$, etc., are shown. If initial data are given for $x = 0$ we could also prescribe data for $x = 0$ and lines would start on the ordinate.

Curves analogous to those of the family C of characteristics can be used for constructing solutions for a whole class of partial differential equations which is called the class of hyperbolic equations. They need not be straight lines.

What is characteristic about characteristics is that along them propagation
of the disturbance, in our case ρ, is governed by an *ordinary* differential
equation, in our case $d\rho/dt = 0$, and that the equation of the characteristic
is also an ordinary differential equation, in our case $dx/dt = c(\rho)$. Thus in
the more general case the solution may be obtained by finding the solution
of two simultaneous ordinary differential equations. This may be done numerically if analytic solutions cannot be found. The use of characteristic curves is
thus a device for transforming partial differential equations into a system of
ordinary ones which can essentially be solved by quadrature.

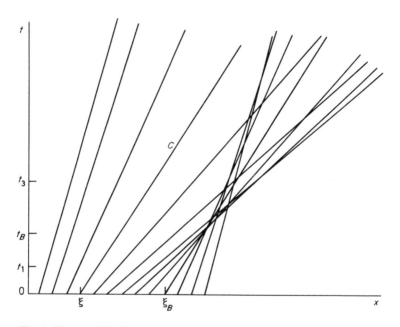

Fig. 1. Characteristic diagram for nonlinear waves. (From [5], with permission.)

Looking at the curves of Fig. 1 we see that for a value of x sufficiently large
there exist several values of ρ, the solution is multivalued. This calls for further
discussion. Each characteristic curve in (x, t) space represents a small part of
the wave, a wavelet, as it were, propagating with its own velocity, i.e. different
parts of the wave travel with different velocities. If wavelets starting further
downstream travel with decreasing velocity, as they do in Fig. 1 they will
eventually be overtaken by wavelets starting upstream. This is shown in Fig. 2
where the distortion of an original waveshape in (ρ, x) space is shown for
different times. Wave shapes at different times may be constructed by moving
a point with a given value of ρ a distance $c(\rho)t$ to the right. It is seen that, since

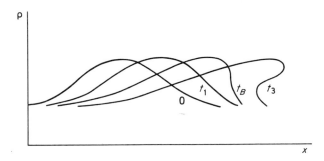

Fig. 2. Breaking wave: successive profiles corresponding to the times 0, t_1, t_B, t_3 in Fig. 1. (From [5], with permission.)

the propagation velocity decreases with x, (a compressive wave) ultimately "breaks" to give a triple-valued solution for $\rho = (x, t)$. Breaking waves can be observed on a sea shore. The breaking starts at $t = t_B$ in Fig. 2 where the profile of ρ develops a vertical (infinite) slope. We can determine the breaking time t_B analytically for our example by calculating ρ_x and finding the time when $\rho_x = \infty$.

From (4)

$$\rho_x = \rho_\xi \xi_x = f'(\xi)\, \xi_x$$

and from (6) $\qquad\qquad\qquad\qquad\qquad\qquad\qquad\qquad\qquad\qquad (7)$

$$\xi_x = 1/(1 + F'(\xi)\, t).$$

Thus if $F'(\xi) < 0$ i.e. if for increasing ξ the velocity $c(f(\xi))$ decreases, the breaking time is given by

$$t_B = -1/F'(\xi). \qquad\qquad (8)$$

Looking at the $x(x, t)$ plane we see the converging characteristics of our compressive wave which must eventually overlap at some time. We can sort out these bundles of lines by observing that a univalued region extends up to a boundary curve which is an envelope of limiting characteristics. (We could find its equation, but we would not learn anything essential from it). From there on there is a bundle of lines with increasing slopes and another with decreasing ones which we could separate out and we could imagine them to lie on different sheets of a folded up (x, t) plane.

These bundles are rather disorderly, so we shall consider a simpler case of a discontinuity in speed due to discontinuous initial data. Let $f(x)$ initially be equal to ρ_1 for $x > 0$ and to ρ_2 for $x < 0$. Accordingly

$$F(x) = \begin{cases} c_1 = c(\rho_1) & x > 0 \\ c_2 = c(\rho_2) & x < 0. \end{cases} \qquad (9)$$

Then two cases may arise, that of an expansive wave with $c_2 < c_1$ or that of a compressive wave with $c_2 > c_1$. The first case is illustrated by Fig. 3 and the second case by Fig. 4.

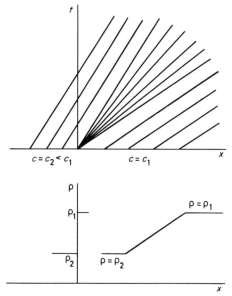

Fig. 3. Centred expansion wave. (From [5], with permission.)

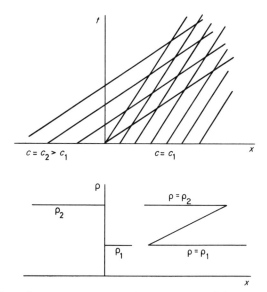

Fig. 4. Centred compression wave with overlap. (From [5], with permission.)

If the initial step function is expansive, as in Fig. 3 a fan of characteristics starts from the discontinuity. One must imagine that all the different densities to which the slopes of the lines correspond occur between the limits ρ_1 and ρ_2. On the other hand in the case of Fig. 4 the breaking occurs immediately. A multivalued region starts right at the origin. Physically this cannot happen and the explanation of the paradox is that some essential physics has not been included in the theory. Let us examine what has been assumed about the mass flow

$$q = \rho v. \tag{10}$$

Assuming that the material (or state) is conserved, the rate of change of the amount contained in a section $x_1 > x > x_2$ must be balanced by a net inflow across x_1 and x_2. Therefore we must have

$$\frac{d}{dt} \int_{x_1}^{x_2} \rho(x, t)\, dx - q(x_1 t) + q(x_2, t) = 0 \tag{11}$$

for continuous derivatives of $\rho(x, t)$ we can take the limit $x_1 \to x_2$ and obtain the conservation equation

$$\frac{\partial \rho}{\partial t} + \frac{\partial q}{\partial x} = 0. \tag{12}$$

If the flow depends on the density alone, i.e. if

$$q = Q(\rho) \tag{13}$$

we obtain

$$\rho_t + c(\rho)\rho_x = 0, \qquad c = Q'(\rho). \tag{14}$$

On the other hand it may be a better approximation to assume that the flow also depends on the density gradient so that, e.g.

$$q = Q(\rho) - v\rho_x \tag{15}$$

even if the correction is small (v small) at breaking point ρ_x becomes large and the term is essential. From (12) and (15) we obtain

$$\rho_t + c(\rho)\rho_x = v\rho_{xx}, \qquad c(\rho) = Q'(\rho) \tag{16}$$

where the term $c(\rho)\rho_x$ leads to breaking as we have seen. The term $v\rho_{xx}$ however introduces diffusion as one knows from the heat diffusion equation

$$\rho_t = v\rho_{xx}.$$

For such an equation the solution of the initial step function problem

$$\rho = \rho_1 \qquad x > 0$$

$$\rho = \rho_2 \qquad x < 0$$

is given by

$$\rho = \rho_2 + \frac{\rho_1 - \rho_2}{\sqrt{\pi}} \int_{-\infty}^{x/(4vt)^{1/2}} e^{-\xi^2} d\xi$$

a smoothed out step approaching values ρ_1, ρ_2 as $x \to \pm \infty$ and with a slope decreasing like $(vt)^{-1/2}$. For small v we approach the step. If (16) is multiplied by $c'(\rho)$ it may be written

$$c_t + cc_x = vc'(\rho)\rho_{xx}. \tag{17}$$

If Q is a quadratic in ρ, $c(\rho)$ is linear in ρ and $c''(\rho) = 0$. We then have

$$c_t + cc_x = vc_{xx} \tag{18}$$

which is known as Burger's equation [6] which has the advantage that it can be solved explicitly and, although it may not completely represent an actual situation gives an insight into the problem of a discontinuity by allowing inspection of the transition of the continuous to the discontinuous solution.

We shall show, that the Burger's equation produces a continuous solution which, in the limit $v \to 0$ yields a discontinuity moving with a certain velocity which we shall denote by U. The discontinuity is a shock which replaces the crosshatched multivalued solution of Fig. 4. The Physics which is missed out in shock wave theory is heat conduction and viscosity which would, when considered in detail, show the shock structure, which is, in our example, a very rapidly varying density.

Let us therefore look for a steady profile $\rho(X)$ which moves with velocity U so that $X = x - Ut$. Then, from (16)

$$(c(\rho) - U)\rho_X = v\rho_{XX} \tag{19}$$

and integrating once we have

$$Q(\rho) - U\rho + A = v\rho_X \tag{20}$$

where A is a constant. Now we are looking for a solution which tends to ρ_1 when $X \to \infty$ and to ρ_2 when $X \to -\infty$ and whose slope ρ_X vanishes at infinity. Thus U and A must satisfy

$$Q(\rho_1) - U\rho_1 + A = Q(\rho_2) - U\rho_2 + A = 0$$

so that U is given by

$$U = \frac{Q(\rho_2) - Q(\rho_1)}{\rho_2 - \rho_1}. \tag{20a}$$

Integrating (20) once more we obtain

$$\frac{X}{v} = \int \frac{d\rho}{Q(\rho) - U\rho + A}. \tag{21}$$

In the case when Q is quadratic in ρ we can write

$$Q - U\rho + A = \alpha(\rho - \rho_1)(\rho - \rho_2) \tag{22}$$

and the integral is easily evaluated. When $c' > 0$, $Q'' > 0$, $\alpha > 0$. We find

$$\frac{X}{v} = \frac{1}{\alpha(\rho_2 - \rho_1)} \log \frac{\rho_2 - \rho}{\rho - \rho_1} \quad \text{or} \quad \frac{\rho - \rho_1}{\rho_2 - \rho} = \exp[-(X/v)\alpha(\rho_2 - \rho_1)] \tag{21a}$$

as $X \to \infty$, $\rho \to \rho_1$, and as $X \to -\infty$, $\rho \to \rho_2$ exponentially.

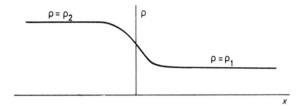

Fig. 5. Shock structure.

We shall have a density change as illustrated in Fig. 5. It is seen that the thickness of the transition region depends on

$$v/\alpha(\rho_2 - \rho_1)$$

and when this is small compared with other typical lengths in the problem the rapid shock transition is well approximated by a discontinuity.

Shock Fitting

We now go back to Fig. 2 showing how the density curve of a breaking wave becomes triple-valued and consider how a discontinuity must be fitted in. It should move with the velocity U of a steady profile given by (20a). How are we to fit this into the continuous solution

$$c = F(\xi) \tag{5}$$

$$x = \xi + F(\xi)t \quad ? \tag{6}$$

Let us consider the triple-valued curve illustrated in Fig. 6. We have replaced this by a discontinuity at position $s(t)$ which cuts off lobes of equal area of the curve. This must be correct since mass and therefore $\int \rho \, dx$ is conserved. While this is true in general, in order to do some analytic work we consider again the case when Q is quadratic in ρ. This is certainly a good approximation in the case of a shock with a small density jump, a weak shock. A Taylor

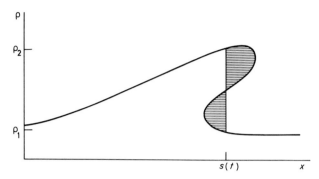

Fig. 6. Equal area construction for the position of the shock in breaking wave. (From [5], with permission.)

expansion in terms of $\rho - \rho_2$ broken off after the second term will yield such a quadratic dependence. In this case we can show another property of the shock. The shock speed is the mean value

$$U = \tfrac{1}{2}(c_1 + c_2) \tag{23}$$

of the speeds $c_1 = c(\rho_1)$ and $c_2 = c(\rho_2)$ on the two sides of the discontinuity. This may be shown by means of (22) by differentiating with respect to ρ. This yields

$$c = Q'(\rho) = \alpha[(\rho - \rho_1) + (\rho - \rho_2)] + U$$

$$c_1 = \alpha(\rho_1 - \rho_2) + U \qquad \text{for} \quad \rho = \rho_1$$

$$c_2 = \alpha(\rho_2 - \rho_1) + U \qquad \text{for} \quad \rho = \rho_2,$$

hence (23). Thus the shock must be fitted in such a way that (23), (5) and (6) are satisfied.

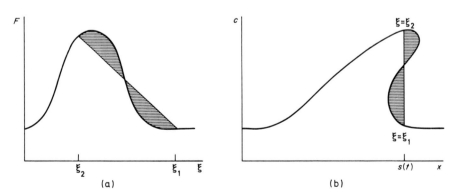

Fig. 7. Equal area construction: (a) on the initial profile; (b) on the transformed breaking profile. (From [5], with permission.)

To look more closely at shock-fitting on the speed curve, consider the initial curve $c = F(\xi)$ at $t = 0$ illustrated in Fig. 7(a), and $c(x)$ at a later time t illustrated by Fig. 7b, which may according to (5) and (6) be obtained from Fig. 7(a) by translating each point a distance $F(\xi)t$ to the right. Since ρ and c are linearly related $\int c\, dx$ is also conserved and the equal area rule has been followed to find the position where the shock is inserted to remove the triple-valued curve. It cuts out the part corresponding to $\xi_2 \geqslant \xi \geqslant \xi_1$.

It may easily be checked that

$$\dot{s}(t) = \tfrac{1}{2}[F(\xi_1) + F(\xi_2)] \tag{24}$$

is obtained when the equal-area rule is obeyed. Mapping the discontinuity of Fig. 7(b) back onto Fig. 7(a), the straight line chord between $\xi = \xi_1$ and $\xi = \xi_2$ of Fig. 7(a) corresponds to the vertical segment on Fig. 7(b) and the equal area property also holds on Fig. 7(a). Characteristics which start out from points ξ_1 and ξ_2 on the x-axis meet at the shock at time t, and the points ξ_1, ξ_2, are any points on the $F(\xi)$ curve such that the chord lops off equal areas. Taking pairs of such points the space time curve of the discontinuity may be constructed and this is illustrated in Fig. 8.

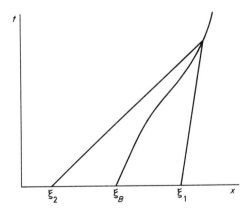

Fig. 8. The (x, t) diagram associated with the shock construction in Fig. 7. (From [5], with permission.)

We are now in a position to indicate a geometric construction applying to more complicated initial data. Suppose initially we have a curve $F(\xi)$ with two humps. Chords of the type Q_1, Q, Q_2 and P_1, P, P_2, shown on Fig. 9(a), will serve to construct two distinct shocks seen in Fig. 9(d). As time proceeds points P_2 and Q_1 approach each other until, as Fig. 9(b) shows, a common chord cuts off lobes of equal area for both humps. The characteristics for Q'_1 and P'_2 are the same and this is the point where, on Fig. 9(d) the shocks have

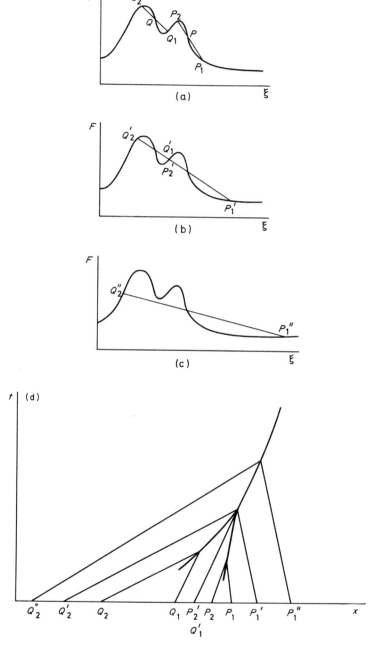

Fig. 9. Construction for merging shocks. (From [5] with permission.)

just merged. The single shock which has now emerged may be constructed by means of a family of chords for which chord P_1'', Q_2'' of Fig. 9(c) is typical. Now only total areas above and below such a chord are made equal for the purpose of the construction.

While the equal area property is a feature of a special assumption about $Q(\rho)$ it is, in general, true that characteristics meet at the discontinuity although it may not be as easy to find pairs that will meet at given times.

Before finishing this section it may be mentioned that the concept of group velocity fits into the general scheme of equation (2). Considering waves with a local wave number $k(x, t)$ and a local frequency $\omega(x, t)$ one might expect that wave crests will be conserved in propagation, so that we have a conservation equation

$$\frac{\partial k}{\partial t} + \frac{\partial \omega}{\partial x} = 0. \tag{25}$$

Hence if k and ω are related by a dispersion equation

$$\omega = \omega(k) \tag{26}$$

we shall have the equation

$$\frac{\partial k}{\partial t} + \omega'(k)\frac{\partial k}{\partial x} = 0 \tag{27}$$

for wave propagation of the local wave number of the "carrier" wavetrain and its propagation velocity is the group velocity

$$v_{gr} = \frac{\partial \omega}{\partial k}. \tag{28}$$

CONSERVATION EQUATIONS AND SHOCK RELATIONS

The conservation equation for the flow

$$\frac{d}{dt}\int_{x_2}^{x_1} \rho(x, t)\,dx + q(x_1, t) - q(x_2, t) = 0 \tag{11}$$

has been stated already. Now it will be shown that it may be used to connect quantities on the two sides of a discontinuity and the shock velocity. It is clear that material cannot pile up at the discontinuity and we shall see that this is avoided when the shock moves with the velocity (20a). Suppose that a discontinuity occurs at $x = s(t)$. What is the function $s(t)$? Let $x_1 > s(t) > x_2$.

To the left and right of the shock ρ and q are continuous and we assume that that they tend to a finite value as $s(t)$ is reached from either side. Then we can write (11) in the form

$$q(x_2, t) - q(x_1, t) = \frac{d}{dt} \int_{x_2}^{s(t)} \rho(x, t)\,dx + \frac{d}{dt} \int_{s(t)}^{x_1} \rho(x, t)\,dx$$

$$= \rho(s^-, t)\,\dot{s} - \rho(s^+, t)\,\dot{s} + \int_{x_2}^{s(t)} \rho_t(x, t)\,dx + \int_{s(t)}^{x_1} \rho_t(x, t)\,dx$$

(following the rule of differentiation of such integrals) where $\rho(s^+, t)$, $\rho(s^-, t)$ are the values of $\rho(x, t)$ as $x \to s(t)$ from above and below respectively, and $\dot{s} = ds/dt$. The integrals tend to zero as $x_1 \to s^+$, $x_2 \to s^-$. Hence

$$q(s^-, t) - q(s^+, t) = \{\rho(s^-, t) - \rho(s^+, t)\}\,\dot{s}. \tag{29}$$

We shall always use subscript 1 for values ahead of shock and subscript 2 for the values behind. Then denoting the shock velocity by $\dot{s} = U$, as before

$$q_2 - q_1 = U(\rho_2 - \rho_1) \tag{30}$$

which may be written in the form

$$-U[\rho] + [q] = 0, \tag{30a}$$

brackets indicating the jumps. This shows a correspondence

$$\frac{\partial}{\partial t} \to -U[\,], \qquad \frac{\partial}{\partial x} \to [\,] \tag{30b}$$

between derivatives and jumps. So if we retain $q = Q(\rho)$ in the continuous parts (30) may be written:

$$U = \frac{Q(\rho_2) - Q(\rho_1)}{\rho_2 - \rho_1}. \tag{20a}$$

A single-valued solution of the problem illustrated by Fig. 11 is possible with a shock moving with velocity U and with

$$\rho = \rho_1 \qquad x > Ut$$

$$\rho = \rho_2 \qquad x < Ut.$$

This solution may be viewed in a frame of reference in which the shock is at rest as illustrated by Fig. 10 or in the framework of the laboratory relative to which the shock moves. The flow quantities are then those of Fig. 11.

Fig. 10. Flow quantities relative to stationary shock.

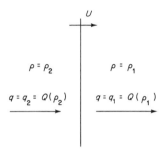

Fig. 11. Flow quantities for moving shock.

We now wish to derive conditions between quantities across a shock for a compressible fluid. We need the conservation equations for the movement of a fluid. We shall state them in three dimensions, denoting the position vector by $\mathbf{x}$ with components x_i, $i = 1, 2, 3$ the density by $\rho(\mathbf{x}, t)$ and the velocity vector has components $v_i(\mathbf{x}, t)$. An external force, e.g. gravity is denoted by components $F_i(\mathbf{x}, t)$ per unit mass. In the case of the macroscopic equations of a plasma of Chapter 4 there are electric and magnetic forces, but the nature of the forces does not concern us here.

The i-component of stress may be written

$$p_i = p_{ji}\lambda_j \tag{31}$$

where λ_i are the components of the unit vector $\hat{e}$ in the direction of the outward normal. In the case of the Stokes–Navier equations

$$p_{ji} = -p\delta_{ji} - \frac{2}{3}\mu\frac{\partial v_k}{\partial x_k}\delta_{ji} + \mu\left(\frac{\partial v_j}{\partial x_i} + \frac{\partial v_i}{\partial x_j}\right) \tag{32}$$

where μ is the viscosity. The first conservation equation

$$\frac{d}{dt}\int_V \rho\,dV + \int_S \rho\lambda_j v_j\,dS = 0 \tag{33}$$

states that total mass change in volume V is balanced by flow across the surface S.

The second one

$$\frac{d}{dt} \int_V \rho v_i \, dV + \int_S (\rho v_i \lambda_j v_j - p_i) \, dS = \int_V \rho F_i \, dV \tag{34}$$

states that the change of momentum in a volume V is balanced by the momentum created by the body forces on the right diminished by the flow of momentum across the boundary represented by the second term on the left.

The third equation

$$\frac{d}{dt} \int_V (\tfrac{1}{2}\rho v_i^2 + \rho e) \, dV + \int_S [(\rho v_i^2 + \rho e) \lambda_j v_j - p_i v_i + \lambda_j q_j] \, dS = \int \rho F_i v_i \, dV \tag{35}$$

states the energy balance. The first term of the volume integral is the change of kinetic energy, the second is the contribution of the internal energy e per unit mass.

The first term in the surface integral arises from transport of energy across the boundary, the second is the rate of working by the stress and the last term is due to the heat flow **q** across the boundary, a vector whose components we have now denoted by q_i (not to be confused with the mass flow in the previous notation). On the right side appears the work done by the external forces.

We have stated these conservation equations in the integral form in which they arise. Conservation equations in differential form need not arise from integral ones. An example is equation (25) and it should be made clear that jump conditions can only be deduced uniquely from those which are the consequence of an integral conservation law.

Written in one dimension, the shock conditions

$$\frac{d}{dt} \int_{x_2}^{x_1} \rho \, dx + \rho v \Big|_{x_2}^{x_2} = 0 \tag{36}$$

$$\frac{d}{dt} \int_{x_2}^{x_1} \rho v \, dx + [\rho v^2 - p_{11}]_{x_2}^{x_1} = 0 \tag{37}$$

$$\frac{d}{dt} \int_{x_2}^{x_1} (\tfrac{1}{2}\rho v^2 + \rho e) \, dx + [(\tfrac{1}{2}\rho v^2 + \rho e) v - p_{11}v + q_1]_{x_2}^{x_1} = 0 \tag{38}$$

follow from (33), (34), (35).

Neglecting viscosity effects ($\mu = 0$ and putting $p_{11} = -p$) and neglecting heat conduction ($q = 0$) we can use the same reasoning which led to the correspondence (30b).

The jump conditions become:

$$-U[\rho] + [\rho v] = 0 \tag{39}$$

$$-U[\rho v] + [\rho v^2 + p] = 0 \tag{40}$$

$$-U[\tfrac{1}{2}\rho v^2 + \rho e] + [\tfrac{1}{2}(\rho v^2 + \rho e) v + pv] = 0. \tag{41}$$

The differential equations holding on each side of the shock follow from (33), (34), (35) because (a) surface integrals may be transformed into volume integrals by using Gauss's theorem; and (b) the equations must be true for arbitrarily small volume elements. They are

$$\rho_t + (\rho v)_x + 0 \tag{42a}$$

$$(\rho v)_t + (\rho v^2 + p)_x = 0 \tag{42b}$$

$$(\tfrac{1}{2}\rho v^2 + \rho e)_t + \{(\tfrac{1}{2}\rho v^2 + \rho e) v + pv\}_x = 0. \tag{42c}$$

In these equations it is assumed, that the internal energy is a state function, i.e. that it is a function of p and ρ or any other pair of thermodynamic variables. These equations may be written more simply by using the derivative

$$\frac{D}{Dt} = \frac{\partial}{\partial t} + v_j \frac{\partial}{\partial x_j} \tag{43}$$

following individual particles. After some manipulation the vector equations become

$$\frac{D\rho}{Dt} + \rho \operatorname{div} \mathbf{v} = 0 \tag{44}$$

$$\rho \frac{D\mathbf{v}}{Dt} + \operatorname{grad} p = \rho \mathbf{F} \tag{45}$$

$$\rho \frac{De}{Dt} + p \operatorname{div} \mathbf{v} = 0 \tag{46}$$

and the last one may be written

$$\frac{De}{Dt} - \frac{p}{\rho^2} \frac{D\rho}{Dt} = 0. \tag{46a}$$

From thermodynamics we know that

$$T \, dS = de + p \, d\left(\frac{1}{\rho}\right)$$

where S is the entropy. Thus (46) becomes

$$T \frac{DS}{Dt} = 0 \qquad (46b)$$

showing that the entropy remains constant following a particle. Such a flow is called adiabatic. It should be noted that the entropy arises here not from thermodynamic equilibrium considerations, the fluid equations just throw up a quantity which figures in equilibrium thermodynamics as the entropy.

Equation (46b) may be replaced by equation

$$\frac{Dp}{Dt} = a^2 \frac{D\rho}{Dt} \qquad a^2 = \left(\frac{\partial p}{\partial \rho}\right)_{S = \text{constant}}. \qquad (47)$$

This follows once the expression for the entropy is solved for p by differentiating p with respect to t and using (46b). Here

$$a = \left[\left(\frac{\partial p}{\partial \rho}\right)_S\right]^{1/2} \qquad (48)$$

is the velocity of sound. For a polytropic gas

$$e = \frac{1}{\gamma - 1}\frac{p}{\rho}, \qquad S = c_v \log \frac{p}{\rho^\gamma}, \qquad a^2 = \frac{\gamma p}{\rho}, \qquad \gamma = \frac{c_p}{c_v}. \qquad (49)$$

If the fluid is initially at rest with uniform entropy, then the entropy is the same on each particle path and remains uniform. Such flows are termed isentropic and p is a function of the density ρ only. For a polytropic gas

$$p = \kappa \rho^\gamma. \qquad (50)$$

The differential equation is only valid in the region where the functions are differentiable. Across a surface of discontinuity the entropy will jump and this jump may change as a function of space and time.

For a polytropic gas equations (39), (40), (41) may be manipulated to yield the shock conditions below which are expressed in terms of a Mach number M defined by

$$M = \frac{U - v_1}{a_1}$$

which is the Mach number of the shock relative to the flow ahead. The relations holding across the shock are:

$$\frac{v_2 - v_1}{a_1} = \frac{2(M^2 - 1)}{(\gamma + 1)M} \tag{51}$$

$$\frac{\rho_2}{\rho_1} = \frac{(\gamma + 1)M^2}{(\gamma - 1)M^2 + 2} \tag{51a}$$

$$\frac{p_2 - p_1}{p_1} = 2\gamma \frac{(M^2 - 1)}{\gamma - 1} \tag{51b}$$

$$\frac{a_2}{a_1} = \frac{\{2\gamma M^2 - (\gamma - 1)\}^{1/2}\{(\gamma - 1)M^2 + 2\}^{1/2}}{(\gamma + 1)M}. \tag{51c}$$

These relations may also be written in another form suitable when p_2 is known. The shock strength

$$z = (p_2 - p_1)/p_1 \tag{52}$$

is introduced and the shock relations take the form

$$M = \frac{U - v_1}{a_1} = \left(1 + \frac{\gamma + 1}{2\gamma} z\right)^{1/2} \tag{52a}$$

$$\frac{v_2 - v_1}{a_1} = \frac{z}{\gamma\left(1 + \dfrac{\gamma + 1}{2\gamma} z\right)^{1/2}} \tag{52b}$$

$$\frac{\rho_2}{\rho_1} = \frac{1 + \dfrac{\gamma + 1}{2\gamma} z}{1 + \dfrac{\gamma - 1}{2\gamma} z} \tag{52c}$$

$$\frac{a_2}{a_1} = \left\{\frac{(1 + z)\left(1 + \dfrac{\gamma - 1}{2\gamma} z\right)}{1 + \dfrac{\gamma + 1}{2\gamma} z}\right\}^{1/2}. \tag{52d}$$

One can then compute the entropy jump which is finite for a finite shock strength z. One finds

$$\frac{S_2 - S_1}{c_v} = \log \frac{(1 + z)\left(1 + \dfrac{\gamma - 1}{2\gamma} z\right)^{\gamma}}{\left(1 + \dfrac{\gamma + 1}{2\gamma} z\right)^{\gamma}}. \tag{52e}$$

G

According to the second law of thermodynamics the entropy must increase when a particle is followed across the shock, thus $S_2 > S_1$. From (52e) $d(S_2 - S_1)/dz > 0$ for $\gamma > 1, z > -1$, which is always true. Hence from $S_2 - S_1 > 0$ we infer that $z > 0$. Thus a shock must always be compressive with $p_2 - p_1 > 0$ and $p_2 > p_1, \rho_2 > \rho_1, a_2 > a_1, v_2 > v_1, M > 1$. This is also a general result in accordance with our previous discussion.

For very strong shocks $z \to \infty$, U is usually large with respect to v_1 and we take $M \approx U/a_1 \gg 1$.

Then we have

$$v_2 \approx \frac{2}{\gamma + 1} U, \qquad \frac{\rho_2}{\rho_1} \approx \frac{\gamma + 1}{\gamma - 1}, \qquad p_2 \approx \frac{2}{\gamma + 1} \rho_1 U^2. \qquad (53)$$

SHOCK SEQUENCES

The strong shock relations (53) show that the density ratio ρ_2/ρ_1, even for infinitely strong plane shocks, is limited to the ratio $(\gamma + 1)/(\gamma - 1)$. Thus if a 1000 or even a 10 000-fold compression is needed for the purpose of laser fusion, a shock sequence must be used. Suppose that a number of successive shocks are applied to a plane target, such that the density ratio for successive shocks is constant and equal to x. The shock strength can then be determined from (52c) and is given by

$$z = \frac{2\gamma(x - 1)}{x(1 - \gamma) + 1 + \gamma}. \qquad (54)$$

From (52) the pressure needed for the first shock p_2 is given in terms of that of the initial material p_1 by

$$p_2 = p_1 + p_1 z$$

and the pressure needed for the nth shock becomes

$$p_n = p_1(1 + z)^{n-1}. \qquad (55)$$

We can now define an effective adiabatic index γ_{eff} by the relation

$$\frac{p_n}{p_1} = \left(\frac{\rho_n}{\rho_1}\right)^{\gamma_{\text{eff}}} \qquad (56)$$

which is given by

$$\gamma_{\text{eff}} = \frac{\log(1 + z)}{\log x},$$

$$x = \frac{\rho_2}{\rho_1} = \frac{\rho_3}{\rho_2} = \dots = \frac{\rho_n}{\rho_{n-1}}. \qquad (57)$$

This effective adiabatic index is never very far from the γ for the material. The following table shows this for $\gamma = \frac{5}{3}$.

TABLE I.

x	3	2·5	2·25	2
γ_{eff}	2·19	1·95	1·87	1·8

The maximum density ratio for $\gamma = \frac{5}{3}$ is $x = 4$. Suppose we want to make $x = 3$: then $z = 10$ and, from (52a) the Mach number is $M = 3$. For a 1000 fold density increase we need $n = 3/\log 3$, i.e. approximately 6·28, say seven shocks. At every step the pressure ratio is 11 so that, starting with 1 bar we need a final pressure of approximately 19 megabars!

The foregoing considerations apply only when shocks do not overtake earlier ones. The nth shock formed at time t_n does not overtake the previous one formed at time t_{n-1} before time t if

$$v_n(t - t_n) \leqslant v_{n-1}(t - t_{n-1}).$$

Treating n as a continuous variable this condition may be rewritten as

$$\frac{dv_n}{dn} \bigg/ \frac{dt_n}{dn} \leqslant \frac{v_n}{t - t_n}$$

which may be integrated to yield

$$\frac{v_n}{v_1} \leqslant \frac{t_1 - t_n}{t - t_n}.$$

All the shocks arrive at time t if the equality sign applies, i.e. if

$$t_n = (v_n t_n - v_1 t_1)/(v_n - v_1)$$

REFLECTION OF A PLANE SHOCK

The reflection of a plane shock from an end wall can easily be treated by means of equation (52). As before, subscripts 1 and 2 refer to the states ahead and behind the incident shock. Let its shock strength be z_i, let subscript 3 refer to the state behind the reflected shock which moves into a region ahead which was left in state 2 by the incident shock. The reflected shock strength is denoted by z_r and is given by

$$z_r = (p_3 - p_2)/p_2. \tag{58}$$

The velocity of the reflected shock however is reversed so that we change the sign and write

$$(v_2 - v_3) = a_2 z_r / \gamma \left(1 + \frac{\gamma + 1}{2\gamma} z_r \right)^{1/2}. \tag{59}$$

Next to the wall the fluid must be at rest and we put $v_3 = 0$. The velocity v_2 is given by

$$v_2 = a_1 z_i / \gamma \left(1 + \frac{\gamma + 1}{2\gamma} z_i \right)^{1/2} \tag{60}$$

from (52b) and a_2 from (52d) by

$$a_2 = a_1 \left[(1 + z_i) \left(1 + \frac{\gamma - 1}{2\gamma} z_i \right) \Big/ \left(1 + \frac{\gamma + 1}{2\gamma} z_i \right) \right]^{1/2}. \tag{61}$$

Thus (59) gives a relation for z_r in terms of z_i

$$\frac{z_i^2}{\gamma_2 (1 + z_i) \left(1 + \frac{\gamma - 1}{2\gamma} z_i \right)} = \frac{z_r^2}{\gamma^2 \left(1 + \frac{\gamma + 1}{2\gamma} z_r \right)}. \tag{62}$$

This leads to a quadratic equation for z_r with the relevant solution

$$z_r = z_i \left(1 + \frac{\gamma - 1}{2\gamma} z_i \right) \tag{63}$$

as may be verified by substitution.

For strong shocks $z_i \to \infty$

$$z_r \approx \frac{2\gamma}{\gamma - 1}, \qquad p_{3/p2} = \frac{3\gamma - 1}{\gamma - 1} = 6 \quad \text{for} \quad \gamma = \tfrac{5}{3}.$$

THE HUGONIOT CURVE

The relation between quantities before and after a shock may be seen at a glance by plotting the locus of the values of p_2 and $1/\rho_2 = \tau_2$, the specific volume, which may be reached from given values p_1 and τ_1 of the undisturbed material in front of the shock. It is shown in Fig. 12.

For a polytropic gas it is a hyperbola with rectangular asymptotes. We show this by first obtaining relations equivalent to (39), (40), (41) in terms of the velocities relative to the moving shock

$$V = (v - U). \tag{64}$$

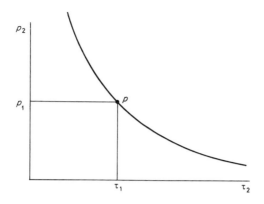

Fig. 12. Values for pressure and specific volume p_2, τ_2 which may be reached from initial values p_1, τ_1.

These relations state simply that the mass flow $m = \rho V$, momentum and energy are conserved, namely

$$m = \rho_1 V_1 = \rho_2 V_2 \tag{65}$$

$$p_0 = p_1 + \rho_1 V_1^2 = p_2 + \rho_2 V_2^2 \tag{66}$$

$$e_0 = e_1 + p_1 \tau_1 + \tfrac{1}{2} V_1^2 = e_2 + p_2 \tau_2 + \tfrac{1}{2} V_2^2 \tag{67}$$

are conserved. Since the enthalpy is defined by

$$h = e + p\tau, \qquad r = \frac{1}{\rho}. \tag{68}$$

Equation (67) can also be written

$$h_0 = h_1 + \tfrac{1}{2} V_1^2 = h_2 + \tfrac{1}{2} V_2^2. \tag{69}$$

From these relations one can easily derive the so-dalled Hugoniot relation

$$e_1 - e_2 = \tfrac{1}{2}(p_2 + p_1)(\tau_2 - \tau_1) \tag{70}$$

which may equivalently be written

$$h_1 - h_2 = \tfrac{1}{2}(p_1 - p_2)(\tau_1 + \tau_2). \tag{70a}$$

This is an equation between state variables on the two sides of the shock.
We may define a Hugoniot function [7]

$$H(\tau, p) = e(\tau_2, p_2) - e(\tau_1, p_1) + \tfrac{1}{2}(\tau_2 - \tau_1)(p_1 + p_2) \tag{71}$$

and the equation of the locus curve

$$H(\tau, p) = 0 \tag{72}$$

for a polytropic gas (71) becomes (using (49))

$$2\mu^2 H(\tau, p) = (\tau_2 - \mu^2\tau_1)p_2 - (\tau_1 - \mu^2\tau_2)p_1,$$

where we have put $\mu^2 = (\gamma - 1)/(\gamma + 1)$.

Thus (72) is indeed the equation of a hyperbola. The three shock relations (65) (66) (67) represent three equations between seven quantities $\tau_1, \tau_2, p_1, p_2, V_1, V_2, U$, since e may be considered given as a function of the state variables p, τ. Hence if three of these parameters are fixed, there is still a one parameter family of possible shocks.

It may be shown in general, that the state on one side of the shock is determined if the state on the other side, and either the shock velocity, or the pressure, or else the velocity on the other side is known. For a polytropic gas, e.g. if v_2 is known the shock strength z may be determined by means of (52b) from which the other quantities may be derived.

In our general context, the processes which occur in reacting cases. i.e. flame propagation, detonations and deflagrations are of interest. They are treated in the same way as a plasma into which, instead of reaction energy, laser light energy is dumped. The equation of state may be different on the two sides.

DETONATION AND DEFLAGRATION

French physicists of the nineteenth century had already noticed that the flame in a tube filled with combustible gas mixtures ignited at one end is propagated with a low velocity of the order of metres per second. This slow combustion may however change over into a very fast process, speeding ahead at some thousands of metres per second which they called detonation. The explanation given by Chapman in 1899 and independently by Jouget in 1905 is in terms of a sharply defined front changing the unburnt gas into burnt gas. The slow regime is usually called a deflagration. The occurrence of these two regimes can be explained by means of the Hugoniot relation once more illustrated in Fig. 13. In the case of endothermic chemical reactions the point p_1, τ_1 lies to the left of the Hugoniot curve. A line drawn from this point may intersect the curve in two places, or it may be tangent to it. We have drawn the tangents at point C, D. The part of the curve in the second quadrant centred at p_1, τ_1, from A onwards corresponds to detonations, the part in the fourth quadrant starting at B to deflagrations. The points C, D, are called Chapman–Jouget points. They have interesting properties. At the detonation point the velocity V_2 of the burnt gas and the entropy S_2 are relative minima, while they are maxima at the deflagration point.

Another property is that the wave speed relative to the burnt gas at these points is equal to its sound speed at these points.

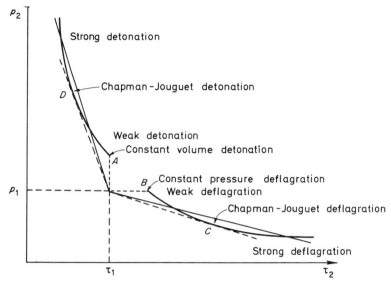

Fig. 13. Hugoniot curve for detonations and deflagrations.

It is easy to show that the entropy is stationary at C and D. The Hugoniot function (71) may be written

$$H^{(2)} = e_2(\tau_2, p_2) - e(\tau_1 p_1) + \tfrac{1}{2}(\tau_2 - \tau_1)(p_2 + p_1) \tag{73}$$

for the burnt gas where, in the case of an exothermic reaction we must include the heat released in e_2. For fixed p_1, τ_1,

$$de_2 = \tfrac{1}{2}(\tau_2 - \tau_1)\, dp_2 + \tfrac{1}{2}(p_2 + p_1)\, d\tau_2 \tag{74}$$

but by definition

$$de_2 = T\, dS_2 - p_2\, d\tau_1 \tag{75}$$

and so

$$dH = T\, dS_2 + \tfrac{1}{2}\{(\tau_2 - \tau_1)\, dp_2 - (p_2 - p_1)\, d\tau_2\}. \tag{76}$$

From (72) it follows that

$$T\, dS_2 = \tfrac{1}{2}\{(\tau_1 - \tau_2)\, dp_2 - (p_1 - p_2)\, d\tau_2\} \tag{77}$$

but for tangency the slope of the curve is equal to the slope of the line drawn from p_1, τ_1 so that

$$\frac{dp_2}{d\tau_2} = -\frac{p_2 - p_1}{\tau_1 - \tau_2} \tag{78}$$

and

$$dS_2 = 0.$$

Now we have learned that at the Chapman–Jouget points $dS = 0$, therefore the adiabatic is tangential to the Hugoniot curve; we also know that the slope $dp_2/d\tau_2$ is related to the sound speed by

$$\rho_2^2 a^2 = -\left(\frac{dp_2}{d\tau_2}\right)_s. \tag{79}$$

Moreover, we find from (65) and (66) that

$$\frac{p_2 - p_1}{\tau_1 - \tau_2} = \rho_2^2 V_2^2 \tag{80}$$

and therefore by (78) the velocity $|V_2|$ is equal to the sound velocity as stated above. This may also be stated by saying that the front, when viewed from the burnt gas, moves with the sound velocity of the burned gas.

So far we have only shown that a detonation may occur at a Chapman–Jouget point. According to Jouget's hypothesis, this is where it actually does occur. This assumption was justified by von Neumann [3] and more recently by Shchelkin et al. [4]. A full discussion including the proof concerning maxima and minima of the entropy etc. may be found in [2] and a full discussion of the chemical processes in explosions by G. I. Taylor in [1].

SIMILARITY SOLUTIONS

In earlier sections it was explained that partial differential equations of fluid flow may be turned into ordinary differential equations along characteristics. There are some problems which are of interest for laser induced compression of matter which can be solved by yet another method. These are the problems of blast waves and implosions. Blast waves are important in diagnostic experiments, while implosions are of course the main theme of the subject. They may be idealised as spherical implosions but the method also works in cylindrical symmetry. For both these symmetries the equations (44) (45) (47) may be written

$$\rho_t + v\rho_r + \rho\left(v_r + \frac{jv}{r}\right) = 0 \tag{81a}$$

$$v_t + vv_r + \frac{1}{\rho}p_r = 0$$

$$p_t + vp_r + a^2(\rho_t + v\rho_r) = 0 \tag{81c}$$

where $j = 1$ for cylindrical and $j = 2$ for spherical symmetry.

Because of the symmetry of the sphere or the cylinder energy injected at the centre producing an explosion is the only parameter on which the solution can depend in addition to the density ρ_0. We shall subsequently denote quantities by ρ, p, v, suppressing the previously used subscript 2 and subscript 0 will replace 1. The total energy will be called E. The dimension of E is ML^2/T^2 and that of ρ_0 is M/L^3. The only parameter involving length and time is E/ρ_0 which has dimension L^5/T^2. Let the coordinate of the shock be a function of time

$$r = R(t) \tag{82}$$

then the only possible form of this time dependence is given by

$$R(t) = k\left(\frac{E}{\rho_0}\right)^{1/5} t^{2/5} \tag{83}$$

where k is a dimensionless number.

We shall use the strong shock relations (53) which yield for pressure and velocity behind the shock

$$p = \frac{8}{25}\frac{k^2\rho_0}{\gamma + 1}\left(\frac{E}{\rho_0}\right)^{2/5} t^{-6/5}, \qquad v = \frac{4}{5}\frac{k}{\gamma + 1}\left(\frac{E}{\rho_0}\right)^{1/5} t^{-3/5} \tag{84}$$

or equivalently

$$p = \frac{8}{25}\frac{k^5}{\gamma + 1} ER^{-3}, \qquad v = \frac{4}{5}\frac{k^{5/2}}{\gamma + 1}\left(\frac{E}{\rho_0}\right)^{1/2} R^{-3/2}. \tag{85}$$

Already some important information has been obtained from a purely dimensional argument. It will turn out that the equations (81a, b, c) can be turned into ordinary differential equations with independent variable

$$\xi = r/R(t) \qquad (\xi = 1 \text{ at the shock}). \tag{86}$$

Any non-dimensional function of r and t can depend only on the combination $\zeta = Et^2/\rho_0 r^5$ and ξ is proportional to $\zeta^{-1/5}$.

The shock being at $\xi = 1$, its velocity U is calculated from (83) to be

$$U = \dot{R} = 2R/5t. \qquad (87)$$

The variables vt/R, ρ/ρ_0, $pt^2/\rho_0 R^2$ are dimensionless. We can write

$$v = n\frac{r}{t}V(\xi), \qquad \rho = \rho_0\Omega(\xi), \qquad p = n^2\left(\frac{r}{t}\right)^2 \rho_0 P(\xi) \qquad (88)$$

where n may be put equal to $\frac{2}{5}$ because of (87). For a point blast, i.e. an explosion starting from a point source we can put $\xi = r/(Ct)^n$ and obtain a solution of the form (88).

Substituting (88) into the equations (81a, b, c), using

$$\frac{\partial}{\partial t} f(\xi) = \xi_t f'(\xi) = -n\frac{\xi}{t} f'(\xi)$$

$$\frac{\partial}{\partial r} f(\xi) = \xi_r f'(\xi) = \frac{\xi}{r} f'(\xi)$$

one finds indeed that the factors r and t cancel out and the following equations in ξ results:

$$\{(V - 1)^2 - A^2\}\xi\frac{\partial V}{\partial \xi} = \left\{(j + 1)V - \frac{2(1 - n)}{n\gamma}\right\}A^2 - V(V - 1)\left(V - \frac{1}{n}\right) \qquad (89a)$$

$$\{(V - 1)^2 - A^2\}\frac{\xi}{A}\frac{dA}{d\xi} = \left\{1 - \frac{1 - n}{\gamma o}(V - 1)^{-1}\right\}A^2 + \frac{\gamma - 1}{2}V\left(V - \frac{1}{n}\right)$$
$$- \frac{\gamma - 1}{2}(j + 1)V(V - 1) - (V - 1)\left(V - \frac{1}{n}\right) \qquad (89b)$$

$$\{(V - 1)^2 - A^2\}\frac{\xi}{\Omega}\frac{d\Omega}{d\xi} = 2\left\{(j + 1)V - \frac{1 - n}{n\gamma}\right\}(V - 1)^{-1}A^2. \qquad (89c)$$

Here we have put $\frac{2}{5} = n$ and $A = (\gamma p/\Omega)^{1/2}$ which is related to the sound speed a by

$$a = n\frac{r}{t}A(\xi). \qquad (90)$$

Analytic solutions to these equations have been calculated by Sedov and his book "Similarity and Dimensional Methods in Mechanics" contains a very full account [8].

GUDERLEY'S IMPLOSION PROBLEM

In the blast wave problem a concentrated energy source drives a shock out to infinity but the implosion problem is of a different nature. It may be understood as an inwardly converging motion driven by a spherically (or cylindrically) contracting piston.

Guderley [9] argues that near the centre of an implosion, r will be given by an expansion

$$r = \sum_m \alpha_m (t_c - t)^m$$

where t_c is the time of collapse and that the smallest m will become dominant for small $t - t_c$. We may assume

$$r = a_m (t_c - t)^\alpha = \text{constant} \left(1 - \frac{t}{t_c} \right)^\alpha \tag{91}$$

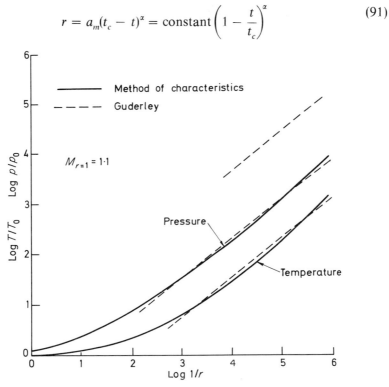

Fig. 14. Theory of converging shock waves. Comparison between sonic theory, Guderley's solution, and the method of characteristics for converging shock waves ($\gamma = 1\cdot40$). p/p_0 and T/T_0 are the ratios of pressure and temperature immediately behind the shock to the initial values. The short upper curve corresponds to the pressure after reflection from the centre. (From [11], with permission.)

where the exponent α and the constant have yet to be determined. Guderley manages to reduce equations (89a, b, c) to a single differential equation and discusses all possible solutions by studying its singularities. He identifies the physically relevant branch of the possible solutions and computes α by numerical integration. For $\gamma = \frac{5}{3}$ he finds $\alpha = 0\cdot717$ a value confirmed by Butler [10] who also computed α for other values of γ both for cylindrical and for spherical implosions. It is remarkable that Fujimoto and Mishkin (F-M) [13] were able to solve the problem analytically by rather elementary means, making only one assumption which appears to be very reasonable. Their solutions will be sketched below.

Guderley also shows that the reflected shock can be fitted into the same similarity solution. He finds that for $\gamma = \frac{7}{5}$ the ratio of the pressure behind the reflected shock to the pressure behind the incident shock is 26 for spherical and about 17 for cylindrical waves. For plane waves the ratio is 8. Stability of the solution is investigated in Whitham's book [5] and an approximate theory shows that small unsymmetrical deviations in the shape of the shock would grow and it would not focus perfectly at the centre, but Whitham believes that instability would affect only a small neighbourhood near the centre.

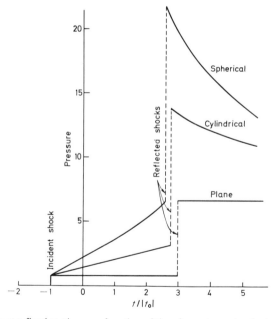

Fig. 15. Pressure at a fixed station as a function of time for a strong implosion. Pressure is measured in units of pressure behind the incident shock and time in units of the time required for the shock to propagate from the fixed station to the centre. The shock arrives at the centre at $t = 0$. (From [11], with permission.)

The flow can also be treated by the method of characteristics and Fig. 14 taken from Perry and Kantrowitz [11] illustrates the agreement between Guderley's theory and the computation. These authors also calculated the reflected shock and Fig. 15 is also taken from their paper. Goldmann [16] has treated the same problem for $\gamma = \frac{5}{3}$ and results for the reflected shock pressure associated density may be found in his paper. He used them for an analytic treatment of neutron yield of an implosion.

We now return to the analytic solution of the implosion problem given by (F-M), which was carried out for an ideal gas with the entropy given by $S = c_v \ln(p\rho^{-\gamma})$. Equations (44), (45) and (46b) may be written in the form

$$D_t\rho + \rho\frac{\partial}{\partial r}v + \frac{2}{r}\rho v = 0 \tag{92a}$$

$$D_t v + \frac{1}{\rho}\frac{\partial}{\partial r}p = 0 \tag{92b}$$

$$D_t(p\rho^{-\gamma}) = 0, \qquad D_t = \frac{\partial}{\partial t} + v\frac{\partial}{\partial r}. \tag{92c}$$

The self similar solutions of these equations are assumed to be separable according to

$$f(r, t) = T(t)\,\Xi(\xi), \qquad \xi = \frac{r}{R} \tag{93}$$

and are expressed in terms of dimensionless functions

$$P(\xi), \qquad \Omega(\xi), \qquad U_1(\xi).$$

It turns out that, by writing U_1 in the form

$$U_1(\xi) = U(\xi) + \xi \tag{94}$$

the equations simplify for a medium with constant ρ_0. Putting

$$p(r, t) = \rho_0\dot{R}^2 P(\xi), \qquad \rho(r, t) = \rho_0\Omega(\xi), \tag{95}$$
$$v(r, t) = \dot{R}[U(\xi) + \xi]$$

they become the ordinary differential equations

$$-\Omega^{-1}\,d_\xi U = U^{-1}\,d_\xi U + 3U^{-1} + 2\xi^{-1} \tag{96a}$$

$$-\Omega^{-1}\,d_\xi P = U\,d_\xi U + (\lambda + 1)\,U + \lambda\xi \tag{96b}$$

$$d_\xi \ln(P\Omega^{-\gamma}) = -2\lambda U^{-1} \qquad \text{or} \tag{96c}$$

$$P^{-1}\,d_\xi P - \gamma\Omega^{-1}\,d_\xi U = -2\lambda U^{-1}. \tag{96d}$$

Here λ is defined by

$$\lambda = R\dot{R}^{-2}\ddot{R} \tag{96e}$$

and it may be shown that $\lambda = $ constant is the condition for separability according to (93). From $\lambda = $ constant follows that

$$R(t) = \text{constant}\left(1 - \frac{t}{t_c}\right)^{\alpha}, \qquad \alpha = \frac{1}{1 - \lambda}. \tag{97}$$

At the shock front $r = R$, $\xi = 1$ and in the case of the strong shock conditions (53) we can write

$$P(1) = \frac{2}{\gamma + 1}, \qquad \Omega(1) = \frac{\gamma + 1}{\gamma - 1}, \qquad U(1) = \frac{1 - \gamma}{1 + \gamma}. \tag{98}$$

The terms of the equations (96a) and (96c) other than those involving U^{-1} integrate and in order to proceed F-M put

$$\sigma(\xi) = \exp\left[-\int_{1}^{\xi} U^{-1}(\xi)\,d\xi\right], \qquad \sigma(1) = 1. \tag{99}$$

It will be seen later that, with the help of an additional assumption the integral may be evaluated. In the case of a strong shock, the result of the integration of (96a, c) from $\xi = 1$ to an arbitrary ξ is

$$P(\xi) = \frac{2}{\gamma + 1}\left[\frac{1 - \gamma}{(1 + \gamma)\,\xi^2 U(\xi)}\right]^{\gamma} \sigma^{2\gamma + 3\gamma} \tag{100a}$$

$$\Omega(\xi) = -\sigma^3/[\xi^2 U(\xi)] \tag{100b}$$

i.e.

$$P(\xi) = \frac{2}{\gamma + 1}\left[\frac{\gamma - 1}{\gamma + 1}\Omega(\xi)\right]^{\gamma} \sigma^{2\lambda}. \tag{100c}$$

Equation (100b) now contains the function U, as yet not known and $\sigma(\xi)$.

The analytic determination of the self-similarity coefficient stems from the following consideration: the derivative of the reduced pressure $d_\xi P$ can be shown from (96b) and (100a) to be positive for $\gamma < 2 + \sqrt{3}$ and for large values of ξ, $P(\xi)$ must decrease. There must be a maximum at some finite value x, y. Setting $d_\xi P = 0$ in (96b) and (96d) leads to

$$xy + (\lambda + 1)\,y + \lambda = 0 \tag{101a}$$

$$\gamma x + 2\gamma y + 2\lambda + 3\gamma = 0 \tag{101b}$$

where

$$x \equiv d_\xi U, \qquad y = \xi^{-1}U. \tag{102}$$

Thus

$$y^2 + \frac{2\lambda + 2\gamma - \gamma\lambda}{2\gamma} y - \frac{\lambda}{2} = 0. \tag{103}$$

If it is now assumed that there is only one point x_m, y_m where $d_\xi P = 0$, it follows that

$$\lambda = \lambda_m = \frac{2\gamma[(8\gamma)^{1/2} - \gamma - 2]}{(2 - \gamma)^2} = -\frac{2\gamma}{(\sqrt{\gamma} + \sqrt{2})^2}. \tag{104}$$

In Table II we compare the value of α_m derived from λ_m with the results of computations of Zeldovich and Raizer ([17], p. 803), Stanyukovich ([18], p. 524) and Brueckner and Jorna ([12] of Chapter 1). (The latter give the value 0·688 for $\gamma = \frac{5}{3}$.) It is seen that the agreement is very good except for $\gamma = 1$.

TABLE II. The Self-similarity Coefficient λ_m and Self-similar Exponent α_m.

γ	1	$\frac{7}{5}$	$\frac{5}{3}$	3	∞
$\lambda_m = \dfrac{2\gamma[(8\gamma)^{1/2} - \gamma - 2]}{(2 - \gamma)^2}$	$-0\cdot343$	$-0\cdot414$	$-0\cdot456$	$-0\cdot605$	$-2\cdot0$
$\alpha_m = \dfrac{1}{1 - \lambda_m}$	$0\cdot744$	$0\cdot707$	$0\cdot687$	$0\cdot623$	$0\cdot333$
Numerically computed values of α_m	$1\cdot0$	$0\cdot717$	$0\cdot688$	$0\cdot638$	$0\cdot375$

F-M also analyse the asymptotic value of the density behind the imploding shock. Equation (100b), remembering the definition (99) of $\sigma(\xi)$ may be written

$$\frac{\rho}{\rho_0} = \Omega(\xi) = -\xi^{-2}U^{-1}(\xi)\sigma^3(\xi) = -U^{-1}\exp\left[-\int_1^\xi \frac{2}{\xi'} + \frac{3}{U(\xi')}\right] \cdot d\xi' \tag{105}$$

since

$$\frac{1}{\xi} = \exp\left(-\int_1^\xi d\xi'/\xi'\right). \tag{106}$$

At the tail of the shock where $\xi \to \infty$ the velocity of the gas is zero and therefore from (9)

$$\lim_{\xi \to \infty} U(\xi) = -\xi \tag{107}$$

and

$$\frac{\rho}{\rho_0} = \Omega(\infty) = \exp\left\{-3\int_1^\infty \left[\frac{1}{\xi'} + \frac{1}{U(\xi')}\right] d\xi'\right\}. \tag{108}$$

In terms of x and y, since differentiation of (102) yields

$$d_\xi x = d_\xi^2 U \, d_\xi y = \xi^{-1}(x - y), \qquad \frac{d\xi}{\xi} = \frac{dy}{x - y}, \qquad (109)$$

the final value of the reduced density is

$$\Omega(\infty) = \exp\left[-3 \int_{y(1)}^{y(\infty)} dy \left(1 + \frac{1}{y}\right) \bigg/ (x - y)\right]$$

$$= \exp\left[-3 \int_{y(1)}^{y(\infty)} \frac{y + 1}{y(x - y)} \, dy\right]. \qquad (110)$$

At $\xi = \infty$ both $x(\infty)$ and $y(\infty) = -1$ and at the front of the shock $x = x(1)$, $y = y(1)$.

The motion of the system behind the shock is illustrated for $\gamma = \frac{5}{3}$ by Fig. 16. The curve starts at the front of the shock and ends at $x(\infty) = y(\infty) = -1$.

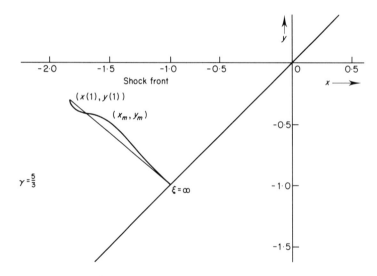

Fig. 16. Track of the shock in the x–y plane.

It passes through the point x_m, y_m mentioned before. Approximating the curved path by a straight line, F-M compute, numerically integrating (110), the following table of final compression ratios.

Figure 17 illustrates the density saturation further. For high γ of an incompressible solid the density ratio is unity. For low γ, the ratio increases very rapidly. The values of Table III do not agree with those of numerical

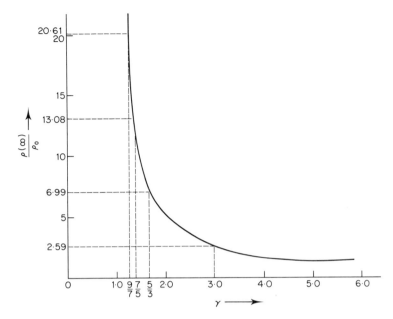

Fig. 17. Density ratios for different values of γ.

computations by other authors, e.g. the value given in [13] for $\gamma = \frac{7}{5}$ is 21·6. It is not clear whether this is due to computational errors in this sensitive range or to the straight line hypothesis of F-M.

TABLE III. The maximum reduced value of the gas density behind the imploding strong shock.

γ	1	$\frac{11}{9}$	$\frac{9}{7}$	$\frac{7}{5}$	$\frac{5}{3}$	3	∞
$x(1)$	$-1\cdot97$	$-1\cdot94$	$-1\cdot93$	$-1\cdot91$	$-1\cdot85$	$-1\cdot59$	$-1\cdot0$
$y(1)$	0	$-\frac{1}{10}$	$-\frac{1}{8}$	$-\frac{1}{6}$	$-\frac{1}{4}$	$-\frac{1}{2}$	$-1\cdot0$
$\Omega(\infty)$	∞	29·31	20·61	13·08	6·99	2·59	1·0

Lagrangian Treatment of Homogeneous Isentropic Compression

Kidder [14] pointed out that there is a shock free isentropic compression scheme which minimizes the compression work and which would thus be preferable to shock sequences. He simplified the analysis by the assumption of homogeneous compression according to which volume elements d^3x change with time everywhere at the same rate, i.e.

$$d^3x = h^3(t)\, d^3x_0. \tag{111}$$

This motion may be looked at as resulting from the shock free merging of particle paths. It may seem rather arbitrary at first, but computer studies show that it is not far from the regime established with the proper initial conditions. It minimizes the compression work, and if it could be established would be preferable to shock sequences. The value of quantities at the origin of time, $t = 0$, will be denoted by subscript 0. We shall treat the compression of a spherical target of density ρ_0 with radius R.

The equations describing the compression are given by (44), (45) and (46b). In spherical polar coordinates with the assumption of angular symmetry the momentum equation (45) may be written

$$\rho \frac{\mathrm{d}v}{\mathrm{d}t} = -\frac{\partial p}{\partial r}. \tag{112}$$

We recall that the equations describe the motion in Lagrangian coordinates, following a given volume element in its movement at successive time intervals $\mathrm{d}t$. In this Lagrangian description each volume element is characterized by the coordinates it had at some reference time which we shall choose to be the origin $t = 0$ of the time coordinate. Thus, in spherical coordinates the position of a fluid element at time t which was at $r = r_0$ at time $t = 0$ will be denoted by $r(r_0, t)$ and according to (111) for homogeneous compression will be given by

$$r(r_0, t) = r_0 h(t) \tag{113a}$$

as a function of time. Similarly

$$v(r_0, t) = \dot{r} = r_0 \dot{h}(t). \tag{113b}$$

The dot indicates time differentiation. The pressure of a fluid element originally at r_0 may, at time t, be denoted by $p(r_0, t)$ and its density by $\rho(r_0, t)$ but we shall also simply write p and ρ for brevity and denote $p(r_0, 0)$ and $\rho(r_0, 0)$ by p_0 and ρ_0 respectively.

In this description an element of mass $\mathrm{d}m_0 = 4\pi\rho_0 r_0^2 \, \mathrm{d}r_0$ originally at r_0 may at a later time be written in the form $\mathrm{d}m = 4\pi\rho r^2 \, \mathrm{d}r$ but since mass is conserved $\mathrm{d}m = \mathrm{d}m_0$ and

$$\rho/\rho_0 = r_0^2/r^2 \, \mathrm{d}r_0/\mathrm{d}r = h^{-3} \tag{114a}$$

so that the equation of motion (112) may be written

$$\dot{v} = -(r^2/\rho_0 r_0^2)(\mathrm{d}p/\mathrm{d}r_0) = r_0 \ddot{h}(t). \tag{114b}$$

We wish to express $\mathrm{d}p/\mathrm{d}r_0$ in terms of $\mathrm{d}p_0/\mathrm{d}r_0$, and we can do this by using the equation of the isentrope which we take to be that of an ideal gas

$$p(r_0, t) = c(r_0) \rho^\gamma(r_0, t). \tag{115}$$

From (106)

$$\rho = \rho_0/h^3(t) \tag{116a}$$

so that

$$p = p_0/h^{3\gamma} \tag{116b}$$

and we obtain

$$\frac{dp}{dr_0} = \frac{dp_0}{dr_0} \, h^{3\gamma}(t). \tag{116c}$$

Using these relations, the equation of motion (114b) becomes

$$h^{3\gamma-2}\ddot{h} = -(1/\rho_0 r_0) \, dp_0/dr_0 \tag{117}$$
$$= -1/t_c^2.$$

The right-hand side of this equation is a function of r_0 alone, while the left-hand side is one of t alone. We recall that r_0 and t are independent Lagrangian variables which now appear separated and this implies that each side must reduce to a constant which we have denoted by $-1/t_c^2$. t_c has the dimension of time and its physical significance will become clear below.

Assuming $\gamma = \frac{5}{3}$ which describes a fully degenerate Fermi electron gas or a nondegenerate fully ionized plasma, a solution of the temporal part of equation (117) is given by

$$h(t) = \left(1 - \frac{t^2}{t_c^2}\right)^{1/2} \quad \text{or} \quad \frac{r^2}{r_0^2} + \frac{t^2}{t_c^2} = 1. \tag{118}$$

This fulfils the initial condition $h(0) = 1$, $\dot{h}(0) = 0$ at $t = 0$, as may be verified by differentiation. Thus the particle time tracks are elliptical. They are shown in Fig. 18.

Now the coordinate $r(R, t)$ of a particle originally at the surface $r = R$ of the pellet will at time $t = t_c$ arrive at $r = 0$ as equations (104) and (113) show. Thus t_c is the time for total collapse. For the pressure at the surface at $r_0 = R$ as a function of time we obtain from (110) and (113)

$$p(R, t) = p(R, 0) \Big/ \left[1 - \left(\frac{t}{t_c}\right)^2\right]^{5/2}. \tag{119}$$

We have thus found that an initial pressure pulse with the time dependence given by (119) is consistent with uniform isentropic compression. In view of the singularity at the centre the formula should not be used too close to $t = t_c$.

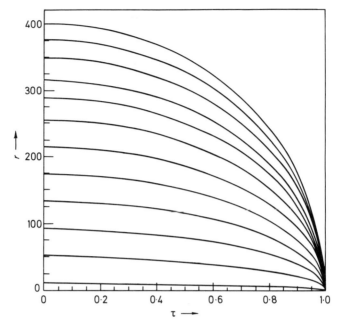

Fig. 18. Homogeneous isentropic compression: radius r (in μm) of fluid particles versus dimensionless time.

A simple, if somewhat artificial assumption concerning the initial spatial distribution of pressure and density leads easily to a solution of the spatial part of (117). Writing (115) for $t = 0$ in the form

$$\frac{p(r_0, 0)}{\rho(r_0, 0)^\gamma} = c(r)_0 \tag{115'}$$

it is seen that the logarithm of the left-hand side is proportional to the entropy. The simple assumption consists in taking it to be independent of r_0. Then the spatial part of (117) becomes

$$c\gamma\rho_0^{\gamma-2} \, d\rho = -\frac{r_0}{t_c^2} \, dr_0 \gamma = c(r_0) \tag{120a}$$

and integrates to

$$c\rho_0^{\gamma-1} = \gamma\frac{p_0}{\rho_0} = \frac{r_0^2}{2t_c^2}(\gamma - 1) + a^2 \tag{120b}$$

where a is a constant determined at $r_0 = 0$ to be the sound propagation velocity

$$a = \left[\gamma\frac{p(0, 0)}{\rho(0, 0)}\right]^{1/2} \tag{120c}$$

at the origin. The pressure distribution is finally given by

$$\frac{p(r_0, t)}{p(0, t)} = \frac{\rho(r_0, t)}{\rho(0, t)}\left[\frac{(\gamma - 1) r_0^2}{2t_c^2 a^2} + 1\right] \tag{121}$$

or, eliminating $\rho(r_0, t)/\rho(0, t)$ by means of (115) with $c(r_0)$ made equal to $c(0)$ (in accordance with our assumption about the initial state)

$$\frac{p(r_0, t)}{p(0, t)} = \left[1 + \frac{(\gamma - 1)r_0^2}{2t_c^2 a^2}\right]^{\gamma/(\gamma - 1)} \tag{122a}$$

$$= \left[1 + \beta\left(\frac{r_0}{R}\right)^2\right]^{5/2} \qquad \text{for} \quad \gamma = \tfrac{5}{3}$$

where we have put

$$\beta = (R/at_c)^2/3. \tag{122b}$$

The density and temperature distributions for $\gamma = \tfrac{5}{3}$ respectively are given by

$$\rho(r_0, t)/\rho(0, t) = \left[1 + \beta\left(\frac{r_0}{R}\right)^2\right]^{3/2} \tag{123a}$$

$$T(r_0, t)/T(0, t) = 1 + \beta\left(\frac{r_0}{R}\right)^2 \tag{123b}$$

When β is small the compression is slow and we see at once that for $\beta \ll 1$ the compressing pellet is almost uniform. For $\beta \gg 1$ however, the collapsing sphere somewhat resembles a spherical shell. The motion can be further studied by the method of characteristics.

TREATMENT OF THE DYNAMICS BY CHARACTERISTICS

We have already explained the use of characteristics in connection with the simple equation (2). It will now be shown that the set (81) of equations can be turned into characteristic form, i.e. into ordinary differential equations holding on characteristic curves. In [5] a general prescription for hyperbolic equations is discussed but in our case the procedure is straightforward: multiply (81a) by a^2, (81b) by $\pm \rho a$ and add the results to equation (81c). The result is that

$$\frac{dp}{dt} \pm \rho a \frac{dv}{dt} + j\frac{\rho a^2 v}{r} = 0 \tag{124}$$

holds on the characteristic curves C^+ and C^- given by

$$C^+: \frac{dr}{dt} = v + a, \qquad C^-: \frac{dr}{dt} = v - a \qquad (125)$$

and

$$\frac{dp}{dt} \quad a^2 \frac{dR}{dt} = 0 \qquad (126)$$

holds on the characteristic $dr/dt = v$. Let us see how this happens, when (81a) multiplied by a^2 (namely

$$a^2(\rho_t + v\rho_r) + a^2\rho\left(v_t + j\frac{v}{r}\right) = 0) \qquad (127)$$

and (81b) multiplied by $\pm \rho a$, (namely

$$\pm a\rho v_t \pm a\rho v v_r \pm ap_r = 0) \qquad (128)$$

is added to (81c) (namely

$$p_t + vp_r - a^2(\rho_t + v\rho_r) = 0) \qquad (129)$$

one obtains

$$p_t + vp_r \pm ap_r \pm \rho a(v_t + vv_r \pm av_r) + j\frac{a^2\rho v}{r} = 0 \qquad (130)$$

but on $C^\pm$

$$dp/dt = p_t + p_r\frac{dr}{dt} = p_t + vp_r \pm ap_r \qquad (131)$$

$$dv/dt = v_t + v_r\frac{dr}{dt} = v_t + vv_r \pm av_r. \qquad (132)$$

For plane flow $j = 0$ and the equations integrate easily. The integration for the spherical symmetry case $j = 2$ for the case of a strong shock carried out by Perry and Kantrowitz has already been referred to. Kidder [14] computed the characteristics on the assumption that the flow is isentropic and homogeneous and with the initial condition of uniform entropy. The pellet radius as a function of t/t_0 is shown in Fig. 19 for a moderate rate of compression. The characteristics represent weak disturbances travelling with the speed of sound. They are seen to arrive at $r = 0$ before the collapse is complete. For rapid compression ($\beta > 10$) they arrive simultaneously at the centre as shown in Fig. 20.

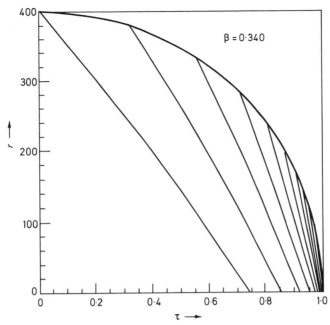

Fig. 19. Homogeneous isentropic compression: radius r (in μm) of pellet surface and ingoing Mach lines versus dimensionless time τ (moderate rate of compression: $B = 0.340$). (From [14], with permission.)

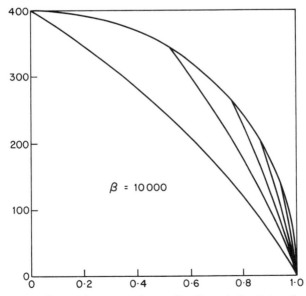

Fig. 20. Homogeneous isentropic compression; radius (in μm) of pellet surface and ingoing Mach lines versus dimensionless time τ (rapid compression: $B = 10$). (From [14], with permission.)

ISENTROPIC COMPRESSION OF THIN SHELLS

As explained in Chapter 2 it may be advantageous to construct fusion targets in the form of spherical shells. In order to apply the treatment of isentropic compression to a shell of outer radius R_s and inner radius R_i we apply the boundary condition $p = 0$ at $r_0 = R_i$. We then obtain from (16)

$$\gamma \frac{p(r_0, 0)}{\rho(r_0, 0)} = \frac{r_0^2 - R_i^2}{3t_s^2} \tag{133}$$

where t_s takes the place of t_0. If we assume that the shell is thin so that $R_s \approx R_i$ and introduce the sound velocity

$$a_s = \left[\frac{\gamma p(R_s, 0)}{\rho(R_s, 0)}\right]^{1/2} \tag{134}$$

at $r_0 = R_s$ and $t = 0$ the constant t_s appropriate for the shell problem is now given by

$$t_s^2 = \frac{2}{3} \frac{R_s \delta R}{a_s^2} \tag{135}$$

where

$$\delta R = R_s - R_i. \tag{136}$$

Density and pressure as a function of r and h are given by

$$\rho(r, t) = \rho(R_s, 0)[(r - R_i)/\delta R]^{3/2} h^{-3} \tag{137}$$

$$p(r, t) = p(R_s, 0)[(r - R_i)/\delta R]^{5/2} h^{-5}. \tag{138}$$

Here r is the radius of a fluid element which was at $r = R_s$ at time $t = 0$. These equations describe the behaviour of a thin shell under the assumption of constant initial entropy. It is of more practical interest to study compression under the assumption of initial uniformity of density and pressure. This has been done by Bevir and the results, which unfortunately need numerical computation of the solution to a differential equation, are reported in a paper by Ashby [15] who also computes the instantaneous power

$$P(r, t) = 4\pi r^2 p(r, t)\left(\frac{\partial r}{\partial t}\right)_R \tag{139}$$

used in compressing the shell.

Ashby reports that computer calculations show that the uniform shell rapidly acquires the profile given by (137). The instantaneous power computed by means of (138), (139), (113a) is given by

$$P = 4 \cdot 25 \left(\frac{e_0}{t_c}\right)\left(\frac{M}{M_0}\right) B^{7/6} t / t_c [(1 - t/t_s)^2]^{-2}. \tag{140}$$

Where e_0 is the internal energy of the shell at $t = 0$, $1/(\gamma - 1)(p_0/\rho_0)$, M_0 is the mass of the shell, M the mass enclosed by radius r. B is the aspect ratio $R_s/\delta R$ of the shell. Here t_c is the collapse time or culmination time R_0/a_0 of a uniform sphere with the same mass, ρ_0 and p_0. This is used to facilitate the comparison between shells and solid spheres.

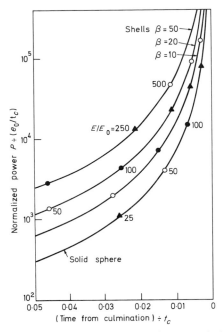

Fig. 21. Mechanical power versus time as culmination is approached for thin shells and solid spheres. [15].

The energy supplied, found by integration of $\dot{P}$ over time is given by

$$E = 1 \cdot 17 \, e_0 B \left(\frac{M}{M_0}\right)\{[(1 - t/t_s)^2]^{-1} - 1\} \tag{141}$$

and the two last equations can be used to relate the peak mechanical power needed to add a certain energy E to a shell by homogeneous compression, namely

$$P = 3\cdot12\left(\frac{e_0}{t_c}\right) B^{-5/6}(E/e_0)^2 [1 + 1\cdot17\,Be_0/E]^{3/2}. \qquad (142)$$

It is interesting to compare power as a function of time near the respective culmination times. Equation (140) becomes

$$P = 0\cdot32(e_0/t_c)\,B^{5/6}\,t_c^2/(t_s - t)^2 \qquad (143)$$

whereas for the solid sphere as $t \to t_0$ one obtains

$$P = 0\cdot727(e_0/t_0)\,t_c^2/(t_c - t)^2 \qquad (144)$$

from Bevirs computations, which yield the same functional relationship. This is plotted in Fig. 21 where the variable on the horizontal axis is the difference between t and the respective culmination points divided by t_c. Mechanical power is plotted versus energy supplied in Fig. 22 where power has been normalized by dividing by e_0/t_c and energy is given in terms of E/e_0. It is seen that for large energy increase, the instantaneous power required is less for shells although it appears from Fig. 21 that the instantaneous power near culmination is higher for shells.

Ashby also shows that for shells a much larger fraction of the energy appears as kinetic energy. Although this kinetic energy can be expected to thermalize at culmination the process then ceases to be isentropic and the entropy increase causes the maximum density to be reduced.

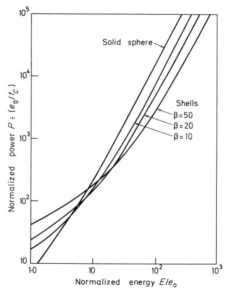

Fig. 22. Mechanical power versus energy supplied for thin shells and a solid sphere. [15].

REFERENCES

1. Emmons, Howard W. (ed.) (1955). "Fundamentals of Gas Dynamics". Oxford University Press, London.
2. Courant, R. R. and Friedrichs. K. O. (1948). "Supersonic Flow and Shock Waves". Interscience, New York.
3. Von Neumann, J. (1942). Office of Scientific Research and Development, Report No. 549. Washington DC.
4. Shchelkin, K. I. and Troshin, Ya. K. (1965). "Gas Dynamics and Combustion" (translated by B. W. Knoshieoff and L. Holtschlag). Mono Bank Corporation, Baltimore.
5. Whitham, G. B. (1974). "Linear and Nonlinear Waves". Wiley Interscience, New York, London, Sydney and Toronto.
6. Burgers, J. M. (1948). *Adv. Appl. Mech.* **1**, 171–199.
7. Hugoniot, H. (1899). *J. l'Ecole Polytech*, **58**, 1–125.
8. Sedov, L. I. (1959). "Similarity and Dimensional Methods in Mechanics". Academic Press, New York and London.
9. Guderley, G. (1924). *Luftfartforschung*, **19**, 302–312.
10. Butler, D. S. (1954). "Converging Spherical and Cylindrical Shocks", Report No. 55/54. Armament Research and Development Establishment, Ministry of Supply, Fort Halstead, Kent.
    Butler, D. S. (1955). "Symposium on Blast and Shock Waves", Report No. 54/54. Armament Research and Development Establishment, Ministry of Supply, Fort Halstead, Kent.
11. Perry, R. W. and Kantrowitz, A. (1951). *J. Appl. Phys.* **22**, 878–886.
12. Zeldovich, Ya. B. and Raizer, Yu. P. (1966). "Physics of Shock Waves and High Temperature Hydrodynamic Phenomena", Vol. I. Academic Press, New York and London.
13. Fujimoto, Y. and Mishkin, E. A. (1977). Private Communication.
14. Kidder, R. E. (1974). *Nucl. Fus.* **14**, 53–60.
15. Ashby, D. E. T. F. (1976). *Nucl. Fus.* **16**, 231–241.
16. Goldman, E. B. (1972). *Plasma Phys.* **15**, 289–310.
17. Zel'dovich, Ya. B. and Reizer, Yu. P. (1967). "Physics of Shock Waves and High Temperature Hydrodynamic Phenomena", Vol. II. Academic Press, New York and London.
18. Stanyukovich, K. P. (1960). "Unsteady Motion of Continuous Media". Academic Press, New York and London.

10

STABILITY

It is well known that magnetic confinement schemes are beset by instability problems. Equilibrium states of plasma confinement are mostly unstable, i.e. small disturbances will grow, so as to disrupt the plasma, and it has proved difficult to take measures to avoid instabilities. One must ask whether analogous problems arise in inertial confinement schemes. The answer is not known at present but there are several well known instability phenomena which may be considered as candidates for the disruptive role which must be examined. A system is stable if any disturbance dies down. On the other hand if any one type of disturbance grows in amplitude, in such a way that the system progressively departs from the initial state and never reverts to it, it is unstable. The stability concept, as stated, applies to stationary states, or equilibrium states. The laser-induced compression however is not a stationary process, and the duration of the process is very short. A disturbance may simply not have time to grow sufficiently to disrupt it. The question we must ask is therefore the following: is the description of the compression in terms of a converging flow correct, or would any departure from shock symmetry or flow symmetry or any other chance disturbance of the flow pattern lead to growing disturbances which would disrupt the flow pattern before compression is achieved?

The problem is a difficult one; the analysis of the compression process under the assumption of spherical symmetry already strains our analytical resources, and computational methods have to be used for realistic predictions. For departures from symmetry 2- or 3-dimensional codes have to be used and it is difficult to get a physical understanding. Intuition may be guided by analysis of simpler, stationary situations. Two phenomena have been analysed in depth which may, *mutatis mutandis*, be encountered in our case, the so called Rayleigh–Taylor instability and the Bénard problem. The former arises when an adverse density gradient exists in accelerated fluid.

To explain what is meant by an adverse density gradient, we consider the simplest case of two fluids of different density which are superposed, one over the other, in the field of gravity. If the density of the upper fluid is higher, the arrangement is clearly unstable. Thus an adverse gradient exists when the density grows in the direction against the accelerating force. The other was experimentally discovered by a French scientist, Bénard, in 1900 [1]. Theoretical foundations for a correct interpretation was laid by Lord Rayleigh in 1900 [2] who also did the fundamental theoretical work on the former instability [3, 4].

Bénard's experiment was concerned with a horizontal layer of fluid in which an adverse temperature gradient is maintained by heating the underside. The gradient is termed adverse since, on account of thermal expansion the fluid at the bottom will be lighter than that above. This top heavy arrangement is again unstable; there will be a tendency for redistribution of the fluid to the extent that its viscosity will allow the motion. It thus turns out that the adverse gradient must exceed a certain critical value before the instability can manifest itself. In the case of the Rayleigh–Taylor instability surface tension at the boundary between layers will stabilize the situation to a certain extent.

In laser induced compression, adverse temperature and density gradients will be encountered in the acceleration field, the former because of the temperature drop behind a thermal front and the latter because the compressed, denser medium is accelerating the uncompressed one.

We shall first proceed with the analysis of the Rayleigh–Taylor problem for an incompressible viscous fluid. The method of the search for instability is to consider small disturbances, and to linearize the equations governing the flow, that is to neglect second order quantities supposed to be small. The disturbances are then assumed to be in the form of normal modes, i.e. they will be developed in series of orthogonal functions forming a complete orthonormal set in terms of which any suitable function can be expanded. We shall then look for modes which are growing in time and examine the growth rates. Fourier components are a suitable orthonormal set and so we shall expand disturbances in terms of functions with space and time dependence given by

$$f(x, y, t) = A \exp(ik_x x + ik_y y + nt). \tag{1}$$

This will allow us to study the growth of disturbances in the direction normal to the z-direction assumed to be that of the (gravitational) acceleration g. It will turn out that the differential equation for the z-dependence will lead to an eigenvalue problem for determining possible values of the growth rate n when the boundary conditions are taken into account. The eigenvalues will be expressed in terms of combinations of the parameters of the problem:

viscosity μ, layer thickness, scale length of density variation $(1/\rho)\,d\rho/dz$ or else the densities of two layers ρ_1 and ρ_2 in the case of a discontinuity. Instability occurs when n is real and positive, but when n is imaginary the fluid is stable.

THE RAYLEIGH–TAYLOR INSTABILITY

We shall only treat the case (a) of two horizontal layers of fluids with densities $\rho_2 > \rho_1$ in a gravitational field of constant acceleration g and (b) the one of a fluid with a density distribution $\rho_0 \exp(\beta z)$. Further cases and details may be found in [5].

We start with equations (44) and (45) of Chapter 9, which were written with co-moving derivatives, but we add to the pressure p, the viscous pressure p_{ik}, (32) of that Chapter. For simplicity we shall deal with incompressible fluids so that

$$\text{div } \mathbf{v} = \frac{\partial v_k}{\partial x_k} = 0 \tag{2}$$

Let the actual density at any point x, y, z be $\rho + \delta\rho$ where $\delta\rho$ is a small density disturbance, let δp be the corresponding pressure disturbance and let the v_i be small. It will be assumed that p and ρ vary in the z-direction only. Neglecting second order quantities, $\partial v_i/\partial t = dv_i/dt$ and the equation of motion (45) becomes

$$\rho \frac{\partial v_i}{\partial t} = -\frac{\partial}{\partial x_i}\delta p + \frac{\partial}{\partial x_k}p_{ik} - g\delta\rho\lambda_i,$$

$$p_{ik} = \mu\left(\frac{\partial v_k}{\partial x_i} + \frac{\partial v_i}{\partial x_k}\right) \tag{3}$$

where $\lambda = (0, 0, 1)$ is a unit vector in the z-direction. In addition we have the equation

$$\frac{\partial}{\partial t}\delta p + v_i\frac{\partial \rho}{\partial x_i} = 0 \tag{4}$$

which expresses that, owing to (2), and (44 of Chapter 9) the density is constant along the particle path.

Differentiating the stress tensor gives

$$\frac{\partial}{\partial x_k}p_{ik} = \left(\frac{\partial v_k}{\partial x_i} + \frac{\partial v_i}{\partial x_k}\right)\frac{d\mu}{dz}\lambda_k + \mu\nabla^2 v_i \tag{5}$$

since

$$\frac{\partial}{\partial x_k}\left(\frac{\partial v_k}{\partial v_i} + \frac{\partial v_i}{\partial x_k}\right) = \frac{\partial}{\partial x_i}\frac{\partial v_k}{\partial x_k} + \sum_k \frac{\partial^2 v_i}{\partial x_k^2} = \nabla^2 v_i \qquad (6)$$

on account of (2), and since μ in first order varies only as a function of z. Hence, writing (3) out in components the three equations

$$\rho\frac{\partial u}{\partial t} = -\frac{\partial}{\partial x}\delta p + \mu\nabla^2 u + \left(\frac{\partial w}{\partial x} + \frac{\partial u}{\partial z}\right)\frac{d\mu}{dz} \qquad (7a)$$

$$\rho\frac{\partial v}{\partial t} = -\frac{\partial}{\partial y}\delta p + \mu\nabla^2 v + \left(\frac{\partial w}{\partial y} + \frac{\partial v}{\partial z}\right)\frac{d\mu}{dz} \qquad (7b)$$

$$\rho\frac{\partial w}{\partial t} = -\frac{\partial}{\partial z}\delta p + \mu\nabla^2 w + 2\frac{\partial w}{\partial z}\frac{d\mu}{dz} - g\delta\rho \qquad (7c)$$

for the components u, v, w of the velocity $\mathbf{v}$ in the x, y and z directions respectively.

Equation (2) becomes

$$\frac{\partial u}{\partial x} + \frac{\partial v}{\partial y} + \frac{\partial w}{\partial z} = 0 \qquad (8)$$

and (4) is now written

$$\frac{\partial}{\partial t}\delta\rho = -w\frac{d\rho}{dz}. \qquad (9)$$

We now analyse the disturbance in terms of the normal modes (1). Equations (7), (8) and (9) become

$$ik_x\delta p = -n\rho u + \mu(D^2 - k^2)u + (D\mu)(ik_x w + Du) \qquad (10)$$

$$ik_y\delta p = -n\rho v + \mu(D^2 - k^2)v + (D\mu)(ik_y w + Dv) \qquad (11)$$

$$D\delta p = -n\rho w + \mu(D^2 - k^2)w + 2(D\mu)(Dw) - g\delta\rho \qquad (12)$$

$$ik_x u + ik_y v = -Dw \qquad (13)$$

$$n\delta\rho = -wD\rho \qquad (14)$$

where

$$k^2 = k_x^2 + k_y^2 \quad \text{and} \quad D = d/dz. \qquad (15)$$

We now want to combine these equations so as to eliminate $\delta\rho$ and δp and obtain an equation for w only. This is accomplished by first multiplying equations (10) and (11) by $-ik_x$ and $-ik_y$ respectively and using (13) to obtain

$$k^2\delta p = [-n\rho + \mu(D^2 - k^2)]Dw + (D\mu)(D^2 + k^2)w \qquad (16)$$

and by combining (12) and (14) to yield

$$D\delta p = -n\rho w + \mu(D^2 - k^2)w + 2(D\mu)(Dw) + \frac{g}{n}(D\rho)w \qquad (17)$$

and finally by eliminating p between equations (16) and (17) so that

$$D\left\{\left[\rho - \frac{\mu}{n}(D^2 - k^2)\right]Dw - \frac{1}{n}(D\mu)(D^2 + k^2)w\right\}$$
$$= k^2\left\{-\frac{g}{n^2}(D\rho)w + \left[\rho - \frac{\mu}{n}(D^2 - k^2)\right]w - \frac{2}{n}(D\mu)(Dw)\right\} \qquad (18)$$

is obtained.

If we suppose that the fluid is confined between two rigid planes the boundary conditions

$$w = 0, \qquad Dw = 0 \qquad (19)$$

must hold on the bounding planes. The latter arises since Dw is related to u and v by equation (13) and these velocity components must vanish. In the case of the two fluids with density ρ_1 and ρ_2 there is a discontinuity at the interface $z = z_s$ and the derivative becomes

$$D\rho = \nabla_s(\rho)\delta(z - z_s), \qquad \Delta_s(\rho) = \rho(z_s + 0) - \rho(z_s - 0) \qquad (20)$$

where $\delta(z - z_s)$ is the Dirac δ-function which is infinite at $z = z_z$ and by definition $\int_{-\infty}^{\infty}\delta(z - z_s)\,dz = 1$. Similarly if μ is discontinuous at $z = z_s$

$$D\mu = \Delta_s(\mu)\,\delta(z - z_s), \qquad \Delta_s(\mu) = \mu(z_s + 0) - \mu(z_s - 0). \qquad (21)$$

On the other hand, w is continuous across the interface, and so are Dw (which can be seen from (13)) and $\mu(D^2 + k^2)w$, the latter because it is easy to show from the definition of the continuous stress tensor and equation (13) that

$$i(k_x p_{xz} + k_y p_{yz}) = -\mu(D^2 + k^2)w \qquad (22)$$

hence equation (18) may be integrated across the interface from $z = z_s - 0$ to $z = z_s + 0$ to yield

$$\left[\Delta_s(\rho) - \frac{1}{n}\Delta_s(\mu)(D^2 - k^2)\right]Dw - \frac{1}{n}[(D^2 + k^2)w]\Delta_s(D\mu)$$
$$= -\frac{k^2}{n^2}g\Delta_s(\rho)w_s - \frac{2k^2}{n}(Dw)_s\Delta_s(\mu) \qquad (23)$$

where w_s and $(Dw)_s$ are the values of these quantities at $z = z_s$ picked up by the δ-function and

$$\Delta_s(D\mu) = (D\mu)_{z+0} - (D\mu)_{z-0}.$$

Equation (23) is a condition which must be satisfied by a solution of case (a) of two uniform fluids of densities ρ_1 and ρ_2 separated by a boundary $z = 0$ perpendicular to g, which we now want to treat.

Case (a).

$$\rho = \rho_2 \quad \text{for} \quad z > 0, \qquad \rho = \rho_1 \quad \text{for} \quad z < 0, \qquad \mu = 0.$$

We shall only treat the case of inviscid fluids ($\mu = 0$) in detail and give some data concerning viscous fluids. For both regions, the general equation (19) reduces to

$$(D^2 - k^2)w = 0 \tag{24}$$

which has the general solution

$$w = A e^{+kz} + B e^{-kz}. \tag{25}$$

Since w must vanish in both z directions, i.e. for $z \to -\infty$ and for $z \to +\infty$ we must take

$$w_1 = A e^{kz} \quad \text{for} \quad z < 0 \tag{26}$$

$$w_2 = A e^{-kz} \quad \text{for} \quad z > 0 \tag{27}$$

and we have put $A = B$ to ensure continuity of w across the interface.

Furthermore we must apply condition (23) which becomes

$$\Delta_0(\rho)Dw + \frac{k^2}{n^2} g(\rho_2 - \rho_1)w_0 = 0 \tag{28}$$

where w_0 is the common value of w at $z = 0$. Applying this condition to the solutions (26) and (27) we obtain

$$k(\rho_2 + \rho_1) = \frac{k^2}{n^2} g(\rho_2 - \rho_1)$$

or $\tag{29}$

$$n^2 = gk \frac{\rho_2 - \rho_1}{\rho_2 + \rho_1}.$$

The last equation shows that the arrangement is stable for $\rho_2 < \rho_1$ since $n^2 < 0$ and unstable for $\rho_2 > \rho_1$ for all wave numbers in the range $k > 0$. This result may have been anticipated by invoking an energy principle. The stable situation must be that of minimum potential energy so that in a

H

gravitational field, the heavier fluid must end up at the bottom, as anybody who has ever turned a glass of water upside down will know. The result does however yield the growth rate of the instability and it shows that the arrangement will not be upset by perturbations of wavelength $\lambda = 2\pi/k$ or smaller during times short compared to

$$\tau = \left(\frac{\lambda}{\pi g}\frac{\rho_2 + \rho_1}{\rho_2 - \rho_1}\right)^{1/2}.$$

In the case of the upturned glass, wavelengths larger than its diameter are of no interest and probably a liquid can be kept longer in an upturned bottle with a narrow neck.

Chandrasekhar shows that a surface tension T changes relation (29) to

$$n^2 = gk\left\{\frac{\rho_2 - \rho_1}{\rho_2 + \rho_1} - \frac{k^2 T}{g(\rho_1 + \rho_2)}\right\}. \tag{30}$$

There is now a maximum wave number

$$k_c = [(\rho_2 - \rho_1)g/T]^{1/2} \tag{31}$$

and the arrangement is stable for sufficiently short wavelengths.

Case (b). We consider an inviscid fluid with a density distribution

$$\rho = \rho_0\,e^{\beta z} \tag{32}$$

in a gravitational field with acceleration g acting in the negative z-direction. In this case equation (19) becomes

$$D^2 w + \beta\,Dw - k^2(1 - g\beta/n^2)\,w = 0 \tag{33}$$

and the boundary condition requires $w = 0$ on the boundaries which we assume to be at $z = 0$ and $z = d$. The general solution is given by

$$w = A_1\,e^{q_1 z} + A_2\,e^{q_2 z} \tag{34}$$

where A_1 and A_2 are arbitrary constants and q_1 and q_2 are the roots of the equation

$$q^2 + q\beta - k^2(1 - g\beta/n^2) = 0. \tag{35}$$

The boundary condition at $z = 0$ is satisfied if

$$w = A(e^{q_1 z} - e^{q_2 z}) \tag{36}$$

while w also vanishes at $z = d$ if

$$\exp[(q_1 - q_2)\,d] = 1 \tag{37}$$

from which it follows that

$$(q_1 - q_2) d = 2im\pi \tag{38}$$

where m is an integer. In conjunction with equation (35) we obtain the result

$$\frac{g\beta}{n^2} = 1 + \frac{\frac{1}{4}\beta^2 d^2 + m^2\pi^2}{k^2 d^2} \tag{39}$$

which shows that $n^2 < 0$ and that the stratification is thus stable for negative β and unstable for positive β, as may have been expected. What is unexpected is that the growth rate shows a maximum as a function of β. The smallest admissible value of m is 1 and this value leads to the largest numerical value of n. For a given set of values of d, k, m, n is numerically largest when β is given by

$$\tfrac{1}{4}\beta^2 d^2 = k^2 d^2 + m^2\pi^2. \tag{40}$$

There is thus a limit on the rapidity of growth, but none on the side of slowness.

We shall treat the case of two uniform viscous fluids separated by a horizontal boundary in less detail, but indicate the manner of approach and results.

Case (c). Fluids of density ρ_1 and ρ_2 and viscosities μ_1 and μ_2 (μ_2 for $z > 0$, μ_1 for $z < 0$). Our equation (19) reduces to

$$D\left[\rho - \frac{\mu}{n}(D^2 - k^2)\right] Dw = k^2\left[\rho - \frac{\mu}{n}(D^2 - k^2)\right] w \tag{41}$$

with $\rho = \rho_1$, $\mu = \mu_1$ or $\rho = \rho_2$, $\mu = \mu_2$ in the two uniform regions. Since ρ and μ are constants we can rewrite the equation in the form

$$\left[1 - \frac{v}{n}(D^2 - k^2)\right](D^2 - k^2) w = 0 \tag{42}$$

where $v = \mu/\rho$ is the coefficient of kinematic viscosity. The general solution of the equation is a linear combination of the two solutions $\exp(\pm kz)$ and $\exp(\pm qz)$ where q is given by

$$q^2 = k^2 + n/v. \tag{43}$$

It is given by

$$w_1 = A_1 e^{+kz} + B_1 e^{q_1 z} \quad \text{for} \quad z < 0, \tag{44a}$$

$$w_2 = A_2 e^{-kz} + B_2 e^{-q_2 z} \quad \text{for} \quad z > 0. \tag{44b}$$

The application of the boundary conditions and the condition at the interface $z = 0$ is more lengthy than in the previous cases, but straightforward. Chandrasekhar shows that n is a function of k and we show this dependence in Fig. 1, for some values of the parameters.

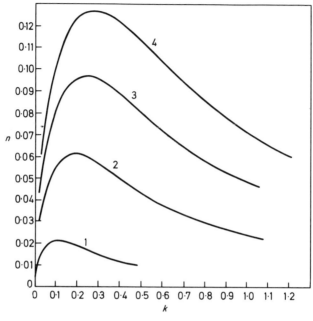

Fig. 1. The dependence of the rate of growth n (measured in the unit $(g^2/v)^{1/3}$) of a disturbance on its wave number k (measured in the unit $(g/v)^{1/3}$ in case the upper fluid is more dense and the kinematic viscosities are the same. The curves labelled 1, 2, 3, and 4 are for values of $(\rho_1 - \rho_2)(\rho_2 + \rho_1) = 0.01, 0.05, 0.10$ and 0.15 respectively. (From [5], with permission.)

TABLE I. The modes of maximum instability for the case $v_2 = v_1$ and $\rho_2 > \rho_1$

$(\rho_2 - \rho_1)/(\rho_2 + \rho_1)$	$k(v^2/g)^{1/3}$	$n(v/g^2)^{1/3}$
0.01	0.1134	0.02081
0.05	0.1939	0.06086
0.10	0.2442	0.09663
0.15	0.2793	0.1267
0.25	0.3304	0.1782
0.50	0.4112	0.2842
0.90	0.4806	0.4265
1.00	0.4907	0.4599

He also gives a table of maximum growth rates and corresponding values of k. In the case of laser fusion, the acceleration is of the order of $5 \cdot 10^{16}$ cm s^{-2} corresponding to a final implosion velocity of $5 \cdot 10^7$ cm s^{-1}. The coefficient v for DT is about 5000 cm s^{-1} at 1 keV and at the solid density of $0 \cdot 19$ g cm^{-3}. From Chandrasekhar's table we find that for large density ratios the maximum values of k and n are given by $k_{max} = 0 \cdot 49 \, (g/v^2)^{1/3}$ and $n_{max} = 0 \cdot 46 \, (g^2/v)^{1/3}$. The corresponding growth rate is about $4 \cdot 10^9$ per s^{-1} at a wave number 600 cm^{-1}. This can seriously affect an implosion unless other effects check the growth.

Brueckner and Jorna [6] report the results of computations which showed that, for the parameter range considered, the ablative flow suppresses the Rayleigh–Taylor instability. Their conclusions are in qualitative agreement with a conjecture due to Nuckolls et al. ([3] of Chapter 2) who found that the effect of the perturbed heat flux on the ablative flow stabilizes the growth of disturbances. Research in this field is still active. Clearly the heat flux must also be considered and we proceed now to sketch the analysis of the Bénard instability and its result.

The Bénard Instability

We have already explained the physical nature of the Bénard phenomenon. It is particularly interesting because it is an example of an instability where the so-called "principle of the exchange of stabilities" applies; this says that the instability sets in as a pattern of cellular convection if, at the onset, a stationary pattern of motion prevails. The striking observation of Bénard was the appearance of cellular flow pattern with a shape reminiscent of a honeycomb, once a certain critical value of a parameter combination was exceeded. The stationary pattern at the onset will be shown to exist to the extent that the growth rate of the perturbation vanishes below and up to this critical value beyond which a negative imaginary part, i.e. growth sets in. Thus the condition for interchange of stabilities is satisfied. The term characterizes the transition from a uniform stationary pattern to a cellular periodic convective regime. The principle may be of wider significance in plasma physics, particularly when magnetic fields are involved as well as electric ones.

In order to analyse the Bénard problem which involves heat flow we must add the balance of the heat flow q, i.e. div q to the energy equation (46 of Chapter 9). On the other hand we shall treat an incompressible fluid and thus put div $v = 0$. The energy equation becomes

$$\rho \frac{De}{Dt} = \text{div } q. \tag{45}$$

Now the internal energy per unit mass may be written as

$$e = \tfrac{1}{2}v^2 + c_v T \tag{46}$$

where c_v is the specific heat at constant volume, T the temperature and q is, by Fourier's law, related to the temperature gradient so that

$$q = K'\nabla T \tag{47}$$

where we have written K' for the coefficient of heat conduction because we shall presently need the coefficient $K = K'/\rho_0 c_v$ of thermometric heat conduction. Thus (45) becomes

$$\rho\left(\frac{1}{2}\frac{\mathrm{D}v^2}{\mathrm{D}t} + \frac{\mathrm{D}c_v T}{\mathrm{D}t}\right) = \nabla . K'\nabla T. \tag{48}$$

Again we shall have perturbations $\delta\rho$ and δp and regard the velocity components as small quantities so that $\mathrm{D}v_i/\mathrm{D}t = \partial v_i/\partial t$. This is why from the outset we have omitted the viscous heat dissipation term which is proportional to the square of the viscous pressure. A more careful consideration shows that it is indeed negligible. Again, we assume an acceleration g to act in the negative z-direction and we can use equation (45) of Chapter 9, augmented by the viscous stress to obtain the equation of motion

$$\rho\frac{\partial \mathbf{v}}{\partial t} = -\nabla p - g\rho\lambda + \mu\nabla^2\mathbf{v} \tag{49}$$

where we have simplified the viscous stress term by means of (2) and (5) and assumed constant μ. λ, as before, has components $(0, 0, 1)$. In the following we shall use the so called Boussinesq approximation which often simplifies the equations. This is concerned with the effects of the expansion law

$$\rho = \rho_0[1 + \alpha(T_0 - T)] \tag{50}$$

where ρ_0 and T_0 are density and temperature at the lower (hot) boundary. The coefficient of expansion is small, and can be ignored, except in the force term because the acceleration resulting from $\delta\rho g = \alpha(T_0 - T)g$ can be quite large.

In the absence of motions (49) simply has the static solution

$$\nabla p = -g\rho_{st}\lambda \tag{51}$$

and we must also have

$$\nabla^2 T = 0 \tag{52}$$

which has the solution

$$T = T_0 - \beta\lambda_i x_i \tag{53}$$

where β is the adverse temperature gradient which is maintained.

If we now consider perturbations and put $p = p_0 + \delta p$

$$T' = T_0 - \beta\lambda_i x_i + \theta, \tag{54}$$

where θ is the temperature perturbation.

The equation of motion (49) becomes

$$\frac{\partial \mathbf{v}}{\partial t} = -\nabla\left(\frac{\delta p}{\rho_0}\right) + g\alpha\theta\lambda + \nu\nabla^2\mathbf{v}, \qquad \nu = \mu/\rho_0 \tag{55}$$

in the Boussinesq approximation. With constant K and c_v the energy equation (48) becomes an equation of heat conduction

$$\frac{\partial \theta}{\partial t} = \beta\lambda_i v_i + K\nabla^2\theta, \qquad K = K'/\rho_0 c_v \tag{56}$$

when only the first order term in v is kept which arises from

$$DT/Dt = \mathbf{v}.\nabla T = -v\beta\lambda.$$

The pressure term can be eliminated from (55). Since the curl of a gradient vanishes, this is accomplished by taking the curl of equation (55). This introduces the quantity $\mathbf{\Omega} = \text{curl}\,\mathbf{v}$, the vorticity. We obtain

$$\frac{\partial \mathbf{\Omega}}{\partial t} = g\alpha\,\text{curl}(\theta\lambda) + \nu\nabla^2\mathbf{\Omega} \tag{57}$$

where the vector $\text{curl}(\theta\lambda)$ is perpendicular to λ. Thus if we take the component along λ we obtain

$$\partial\zeta/\partial t = \nu\nabla^2\zeta \tag{58}$$

an equation for the z-component of $\mathbf{\Omega}$ which we denote by ζ. Taking the curl of equation (57) in turn and taking its z-component we obtain

$$\frac{\partial}{\partial t}\nabla^2 w = g\alpha\left(\frac{\partial^2\theta}{\partial x^2} + \frac{\partial^2\theta}{\partial y^2}\right) + \nu\nabla^4 w \tag{59}$$

where we have denoted the z-component of $\mathbf{v}$ by w.

Our working equations are now (58), (59) and (56), the latter is now written as

$$\frac{\partial \theta}{\partial t} = \beta w + K\nabla^2\theta. \tag{60}$$

As a boundary condition we specify that w and θ must vanish at the bottom $z = 0$ and at the top, $z = d$ of the layer. Further boundary conditions arise which are different for free and rigid boundaries. At a rigid boundary the x and y components u and v of $\mathbf{v}$ must vanish for all x and y, hence, since div $\mathbf{v} = 0$, $dw/dz = 0$. On a free surface the components p_{xz} and p_{yz} of the viscous stress must vanish. Therefore $\partial v/\partial z = \partial v/\partial z = 0$ on a free surface, and one sees by differentiating div $\mathbf{v}$ with respect to z that $d^2w/dz^2 = 0$ on a free surface.

In a way similar to what was done in the section on the Rayleigh–Taylor instability we study disturbances

$$w = W(z)\exp[i(k_x x + k_y y) + pt] \tag{61a}$$

$$\theta = \Theta(z)\exp[i(k_x x + k_y y) + pt] \tag{61b}$$

$$\zeta = Z(z)\exp[i(k_x x + k_y y) + pt] \tag{61c}$$

$$k_x^2 + k_y^2 = k^2. \tag{61d}$$

Introducing these perturbations into equations (58), (59) and (60) we obtain

$$p\left(\frac{d^2}{dz^2} - k^2\right)W = -g\alpha k\Theta + v\left(\frac{d}{dz^2} - k^2\right)^2 W \tag{62a}$$

$$p\Theta = \beta W + K\left(\frac{d}{dz} - k^2\right)\Theta \tag{62b}$$

$$pZ = v\left(\frac{d}{dz^2} - k^2\right)Z \tag{62c}$$

we want solutions of these equations with $\Theta = 0$ and $W = 0$ at both boundaries and satisfying

$$Z = dW/dz = 0 \text{ on a rigid boundary} \tag{63}$$

and

$$dZ/dz = d^2W/dz^2 = 0 \text{ on a free boundary.} \tag{63a}$$

It is convenient at this stage to introduce non-dimensional variables, so that it becomes clear on which parameter combination the onset of instability depends.

We choose units $[L] = d$ and $[T] = d^2/v$ and express k and p in terms of the non-dimensional wave number a and time constant s by

$$k = a/d, \qquad p = s/(d^2/v) \tag{64}$$

we shall not change the notation for the coordinates and let x, y, z stand for the non-dimensional quantities x/d, y/d, z/d. Equations (62a, b) become

$$(D^2 - a^2)(D^2 - a^2 - s) W = \left(\frac{g\alpha}{v} d^2\right) a^2 \Theta \tag{65}$$

$$[D^2 - a^2 - (v/K) s] \Theta = -(\beta d^2/K) W \tag{66}$$

where $D = d/dz$. W and Θ still have their usual dimensions: we do not normalize them with d.

By eliminating Θ between the last two equations we obtain

$$(D^2 - a^2)(D^2 - a^2 - s)[D^2 - a^2 - (v/K) s] W = -Ra^2 W \tag{67}$$

where R is the Rayleigh number, which turns out to govern this problem and is given by

$$R = \frac{g\alpha\beta}{Kv} d^4. \tag{68}$$

It may be shown that s is real for $R > 0$, i.e. for all adverse temperature gradients but we omit the proof which may be found in [5].

This means that at the onset of the instability p and therefore s must go to zero before they can go negative. The marginal state where the system just goes over from stability to instability is stationary (there are no oscillations in time) and the principle of interchange of instability is satisfied. The equation

$$(D^2 - a^2)^3 W = -Ra^2 W \tag{69}$$

for the marginal state is obtained by putting $s = 0$ in equation (67). From (65) and (66) we infer that for $s = 0$ the boundary condition at $z = 0$ and $z = 1$ is $W = 0$ and $(D^2 - a^2) W = 0$ and either DW or $D^2 W = 0$ depending on the nature of the surfaces at $z = 0$ and $z = 1$.

We have a differential equation of order six and six boundary conditions which cannot in general be satisfied. Only for particular values of R will a non-trivial solution exist and we thus have a characteristic value problem before us. If both surfaces are free the boundary value problem is particularly easy to solve. At both surfaces $W = D^2 W = D^4 W = 0$ and from (69) it follows that $D^6 W = 0$. But we can go on and carry out double differentiation indefinitely so that all even derivatives must vanish. Thus the solution must be

$$W = A \sin n\pi z \tag{70}$$

where A is a constant and n is an integer. Equations (69) thus furnishes the characteristic value

$$R = (n^2\pi^2 + a^2)^3/a^2. \tag{71}$$

For given a, the lowest value occurs for $n = 1$. This means that for all wave numbers less than that given by (71) disturbances with wave number a will be stable; that they will be marginally stable when R is given by (71) and when the number R exceeds that value they will be unstable. There is a critical wave number, determined by the maximum of R reached when $\partial R/\partial a^2 = 0$. This happens when $a^2 = \pi^2/2$ and the corresponding value of R is given by

$$R_c = (27/4)\,\pi^4 = 657{\cdot}5.$$

TABLE II. The parameters characterizing the marginal state
for three cases

Nature of the bounding surfaces	R_c	a	$2\pi/a$
Both free	657·511	2·2214	2·828
Both rigid	1707·762	3·117	2·016
One rigid and one free	1100·65	2·682	2·342

Chandrasekhar gives a table of results for other boundary conditions. For the typical laser fusion data we can estimate a critical thickness of the layer beyond which the instability sets in. Setting $\beta/T = (1/T)\,dT/dz = 1/d$ we obtain

$$R_c = 10^9\,d^3$$

where d is measured in cm. Thus $d \approx 10^{-2}$ cm which is considerably larger than the distance over which the temperature of the thermal front develops. Again, Brueckner et al. ([12] of Chapter 2) conclude from numerical computations including flow, that there is no instability. Such conclusions may not be valid if, due to non-uniform illumination, the pressure and temperature distribution is non-uniform to start with.

Both the Rayleigh and Bénard problem should be investigated in the appropriate spherical flow conditions of laser compression. Chandrasekhar thinks that elementary methods of solution are impracticable, but he does treat the problem of the oscillations of a viscous liquid globe, originally solved by Lord Kelvin.

We now briefly discuss the cell patterns which develop when the Bénard problem is studied beyond the marginal state. This pattern is not specified without symmetry conditions imposed in addition to the temperature gradient. Essentially this is so because a given wave vector can be resolved into two orthogonal components in infinitely many ways. Possible symmetries require triangular, square, or hexagonal patterns.

A square pattern arises from the solution

$$w = W(z) \, w \frac{2\pi}{L_x} x \cos \frac{2\pi}{L_y} y.$$

Figure 2 illustrates the infinite sequence of flow cells by showing three of them with the streamlines of the flow. Christopherson [7] discovered the solution

$$w = \tfrac{1}{3} W(z) \{ 2 \cos(2\pi x/L\sqrt{3}) \cos(2\pi y/3L) + \cos(4\pi y/3L) \}$$

for the hexagonal pattern which appears to have been observed by Bénard.

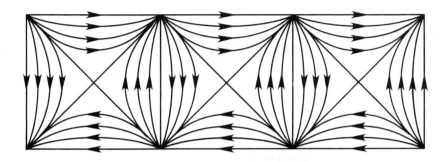

Fig. 2. Convection cells in the case of square symmetry.

REFERENCES

1. Bénard, H. (1901). *Revue générale des Sciences pures et appliquées,* **11**, 1261–71; *ibid.* 1309–1328.
2. Lord Rayleigh (1916). *Phil. Mag.* **32**, 329–346.
3. Lord Rayleigh (1900). Scientific Papers, 200–207. Cambridge, England.
4. Taylor, G. I. (1950) *Proc. Roy. Soc. (London)* A **201**, 192–96.
5. Chandrasekhar, I. (1961). "Hydrodynamic and Hydromagnetic Stability". Oxford University Press, London.
6. Brueckner, K. A., Jorna, S. and Janda, R. (1974). *Phys. Fluids,* **17**, 1554–1559.
7. Christopherson, D. G. (1940). *Quart. J. Math., Oxford,* **11**, 63–65.

11

PLASMA AT HIGH DENSITY, PRESSURE AND TEMPERATURE

In this chapter we review the existing knowledge on the properties of plasma which are encountered along the fusion track, i.e. for temperatures up to 50 kV, for densities up to 10^4 times solid density and corresponding pressures. This review will not be confined to hydrogen and its isotopes, in view of research which may become possible, by means of laser fusion facilities, on interesting properties of other materials. There are still very important gaps in our knowledge.

At high pressures and temperatures the behaviour of matter is not simply that of a gas, liquid or solid. The state of matter depends on the interaction between electrons, ions (or nuclei when the ions are completely stripped of electrons).

The properties of a plasma in more extreme parameter regions are of great interest in astrophysics but laser fusion plasma occupies only a small corner in parameter space. Already in 1936, Hund reviewed the astrophysically important regions of this space and his results, summarized in Fig. 1 may still serve for a general orientation.

In a classical plasma with Coulomb interaction, all thermodynamic quantities depend on Γ only and not separately on density and temperature. Unfortunately, when quantum effects have to be taken into account this is no longer so.

In the case of an electron gas, Montroll and Ward [18] argue, that in the general case the only non-dimensional combination of the density $n = N/V$, h, m, and $\beta = 1/kT$ is $\beta h^2 n^{2/3}/m$. At low temperatures, the ratio

$$e^2 n^{1/3} \left/ \frac{h^2}{m} n^{2/3} \right. = me^2/h^2 n^{1/3}$$

of the potential and the Fermi-energy is appropriate. Thus the thermodynamic potential or partition function should depend on these quantities like

$$me^2 n^{-1/3} h^{-2} f(\beta h^2 n^{2/3} m^{-1})$$

where $f(x)$ tends to unity as $\beta \to \infty$ and to x as $\beta \to 0$. This has the effect that it depends on Γ for high temperatures and at the low limit on the combination $me^2/h^2 n^{1/3}$.

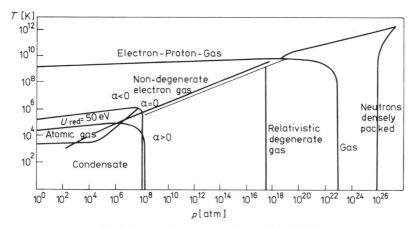

Fig. 1. States of matter according to Hund [11].

Typical values for the quantities involved which are of interest for laser driven fusion are listed in Table I below.

It can be seen that the thermal De Broglie wavelength is always larger than the Debye shielding distance so that quantum effects must be expected. We shall see below that the degree of degeneracy may be characterized by the number of electrons in a sphere with radius λ_T. We shall also see that Γ values of 10–50 lead to important ion configuration effects.

The region labelled "condensate" may be further subdivided and one may wish to know whether we have a liquid, a solid, and, in particular a molecular or metallic solid or liquid. We shall deal in some detail with degeneracy, a

TABLE I

	$T = 10\,\text{keV}$ $n = 10^{31}/\text{m}^3$	$T = 1\,\text{eV}$ $n = 10^{31}/\text{m}^3$	$T = 1\,\text{eV}$ $n = 10^{29}/\text{m}^3$
λ_D	$7\cdot43 \cdot 10^{-10}$	$7\cdot43 \cdot 10^{-9}$	$7\cdot34 \cdot 10^{-8}$
λ_T	$1\cdot22 \cdot 10^{-9}$	$1\cdot22 \cdot 10^{-7}$	$1\cdot22 \cdot 10^{-7}$
Γ/Z^2	$5 \cdot 10^{-3}$	50	$10\cdot7$

consequence of Fermi–Dirac statistics for electrons, according to which the distribution of electron density n as a function of temperature is not a Maxwellian, but follows the Fermi-law

$$n_i \approx \frac{1}{\exp(-\alpha + E_i/kT) + 1} \tag{1}$$

where n_i are the electron numbers in energy level E_i (in equilibrium) and where α is called the degeneracy parameter because it governs the deviation of the Fermi-gas properties from those of a classical gas. In this diagram, $\alpha = 0$ is taken as the dividing line between classical and degenerate behaviour. Fig. 1 leaves out other important subdivisions of parameter space. For one, degeneracy is not the only quantum effect to be considered: there are diffraction effects as well which are a consequence of the wave-nature of electrons and they are not characterized by the parameter α. What matters for these effects is ratio of the De-Broglie wave-length λ_T of a thermal electron $h/(2mkT)^{1/2}$ to the Debye length λ_D which characterizes the range of the force due to ions immersed in an electron fluid.

Another non-dimensional number is of great importance for a Coulomb gas in the absence of quantum effects, the ratio of the potential $(Ze)^2/4\pi a \varepsilon_0$ between ion pairs and their kinetic energy. Here a is usually taken to be the "ion radius", $a = [3/(4\pi n)]^{1/3}$. The stronger the Coulomb interaction is compared to the thermal energy, the more will be the tendency to condensation. Moreover we shall see later, that the boundary between the liquid and crystalline state depends critically on this ratio

$$\Gamma = (Ze)^2/4akT\pi\varepsilon_0 \tag{2}$$

THERMODYNAMIC PROPERTIES

The properties of matter at high density, temperature and pressure (high dtp) to be investigated are those which are important for the compression process. Clearly the pressure as a function of temperature and density is crucial. The specific heat enters into the calculation of the shock wave propagation in the target. It is also of interest to see whether there are other features of the behaviour of high dtp matter which might be experimentally investigated with the help of a laser fusion facility, namely phase transformation, and the production of interesting compounds which can only be formed in the parameter regime available.

Thermodynamics establishes the free energy F

$$F = U - TS \tag{3}$$

(U internal energy, T temperature, S entropy)

as the key concept, the quantity from which all others may be derived. The system may depend on the configurational coordinates $q: q_1, q_2, q_3, \ldots$ Of the electrons and ions and also on external forces through the Hamiltonian, $H(q, p, a)$, i.e. the energy as a function of the momenta, internal coordinates q and the external coordinates $a: a_1, a_2, \ldots$, such as volume, pressure or external fields. The external forces may be defined by

$$A_i = -\frac{\partial H}{\partial q_i}(q, p, a). \tag{4}$$

A change in the free energy dF is then related to these variables by

$$dF(T, a) = (F - U)\, dT/T - \sum_i A_i \, da_i. \tag{5}$$

The system may contain different constituents, say different ion species with number densities $N_1, N_2, N_3, \ldots$. In this case dF becomes

$$dF(T, N, a) = (F - U)\, dT/T - \sum_j \mu_j \, dN_j - \sum_j A_i \, da_i \tag{6}$$

where μ_j, the chemical potentials measure the change in free energy brought about by changes of the different number densities.

Looking at the problem from the point of view of statistical Physics, it turns out that the free energy is known as soon as the partition function for an ensemble of particles

$$Z(\beta, N, a) = \frac{1}{N! h^{3N}} \int \exp[-\beta H(q, p, a)]\, dq\, dp \tag{7}$$

(where $\beta = 1/kT$) is defined. The connection between the statistical definition and the thermodynamic one is complete if the internal energy is the average $\langle H \rangle$ of the particle Hamiltonian and similarly the forces A_i are the averages $\langle A_i \rangle$ of the forces acting on particles.

The connection between the free energy and the partition function is given by

$$F = -\beta^{-1} \ln Z \tag{8}$$

and the chemical potentials are given by ([17], p. 17 f.f.)

$$\mu = -\beta^{-1} \frac{\partial}{\partial N} \ln Z. \tag{9}$$

The entropy becomes

$$S = -\frac{\partial F}{\partial T} = \frac{\partial}{\partial T} kT \ln Z(T, N, V). \tag{10}$$

The internal energy is

$$U = F + TS = kT^2 \frac{\partial}{\partial T} \ln Z(T, N, V). \tag{11}$$

The specific heat is given by

$$C_V = \frac{\partial U(T, N, V)}{\partial T} = \frac{\partial}{\partial T} \left[kT^2 \frac{\partial}{\partial T} \ln Z(T, N, V) \right], \tag{12}$$

and the pressure by

$$P = -\frac{\partial F}{\partial V} = kT \frac{\partial}{\partial V} \ln Z(T, N, V). \tag{13}$$

All this is explained in the books on statistical mechanics, e.g. Landau and Lifshitz (1959).

The quantum mechanical formulation according to which

$$Z = \sum_j e^{-\beta E_j}, \tag{14}$$

where the E_j are the energy eigenvalues of the system, shows particularly clearly how the partition function summarizes all the relevant energetic information pertaining to the system. It is almost intuitively obvious that properties must depend on the ratio of the energy levels to the thermal energy and that the function whose logarithm is proportional to the free energy must be an additive function of functions of these energy ratios.

The partition function can also be written in the form

$$Z = \int \sum_j e^{-\beta E_j} \psi_j^*(r) \psi_j(r) \, d^{3N} r = \int \psi_j^*(r) e^{-\beta H} \psi_j(r) \, d^{3N} r, \tag{15}$$

where the ψ_j are the (orthonormal) wave functions associated with the energy levels. To see this, it is only necessary to write the series (symbolically, since the exponent contains an operator) for the exponentials which then involves powers of H. But $H\psi_j = E_j\psi_j$ and the total probability $\psi_j^*\psi_j$ normalizes to unity, i.e.

$$\int \psi_j^* \psi_j \, d^{3N} r = 1. \tag{16}$$

There is a vast literature concerning the computation of the partition function, and moreover one where a great deal of mathematical ingenuity has been used to good advantage. For low values of Γ the results are well known and are originally due to Debye and Hückel [13]. They apply to the theory of strong electrolytes. The effect of the electrons on the ion interaction consists in shielding the Coulomb potential. We cannot attempt to provide a full bibliography of the subject but the "Les Houches Lectures" (1959) contains among other notable contributions a very readable account written by E. W. Montroll [17]. Unfortunately, even the most advanced analytic work does not yield quantitative information for our purpose but within the last few years very accurate results have been obtained by numerical methods, by de Witt *et al.* and by Hansen and his collaborators [4, 9].

The method of these computations goes back to a pioneering paper by Brush *et al.* [1] in which the Monte Carlo method is applied to the problem. The model treated consists of a system of identical point charges immersed in a uniform background. The system is made electrically neutral by choosing the continuous charge density of the background equal and opposite to the average density of the point charges. The positive point charges represent the charged nuclei of the elements considered and the background represents approximately the effect of the free electrons. The authors quoted deal with cubic cells containing up to 500 particles. The potential energy of interaction between particles is the Coulomb potential $\phi(r) = (Ze)^2/r\varepsilon_0$. The cubic cells are assumed to be periodically repeated in the three directions of space. Therefore the potential must be modified to take account of the interaction with particles in other cells. A charged particle in a cell has a three-dimensional lattice of image charges in the other cells and so, in order to calculate the potential energy of ion pairs, the interaction of an ion with the background and all the images of the other particles has to be computed. This is greatly facilitated by a method due to Ewald, well known in solid state physics, which replaces the lattice sums by fast converging series.

Since the calculation is a classical one, all quantities are given in terms of one non-dimensional parameter Γ, defined above, so that the ratio of the potential energy and kT, ϕ/kT is given by Γ/x where $x = r/a$.

The potential due to a charge distribution $\phi(r)$ is given by a solution of Poisson's equation:

$$\varepsilon_0 \nabla^2 \phi = -eN(r). \tag{17}$$

Thus it is seen that a uniform background of charge does not simply add a constant, but modifies the spatial variation of the potential. As a matter of fact, it may be shown that without the uniform background the configuration would be unstable.

With such a background, one would expect the model to be adequate at very high densities and low temperature when the system becomes pressure-ionized and the degenerate electrons have sufficient energy to have a nearly uniform distribution. On the other hand, at moderate densities and high temperatures the system becomes temperature-ionized. The model is again adequate if the free electrons possess sufficient energy to be uniformly distributed. A model with non-uniform distribution describing certain aspects of a regime where quantum effects are important will be discussed below.

The application of the Monte Carlo method, originally introduced by Metropolis *et al.* [6] (see also [8]) will now be briefly explained. First an initial configuration is chosen; this means that the coordinates (x, y, z) of N particles in the basic cell are specified. (The velocities need not be specified since the force law is velocity independent and velocity space integrations can be done analytically). For an initial particle distribution positions may be selected randomly, or else arranged according to a regular lattice. The computer then calculates the potential energy from an analytic lattice energy formula. Secondly, the computer chooses randomly a number between 1 and N to designate a particle which it will attempt to move. A trial move is constructed by choosing displacements in the x, y and z directions, each of which may vary randomly between Δ and $-\Delta$. If the trial move carries the particle out of the cell, it is brought back, i.e. replaced by its image. Δ is chosen by experience so as to achieve results independent on this choice for very long runs. Thirdly, the computer must decide whether to accept the trial move. According to a preferred method it is always accepted if it lowers the energy. It is accepted with a probability $\exp(-\Delta\varepsilon/kT)$ (Boltzmann factor) if it raises the energy by $\Delta\varepsilon$.

Using a chain of configurations constructed in this way (a Markov chain since each move is independent of previous ones) any function of configuration can be obtained by averaging over all members of the chain. The limiting frequency of each configuration will then be proportional to the corresponding Boltzmann factor. It turns out that the equilibrium value of the pair interaction energy, or pair potential, is approached after a few thousand configurations have been calculated. In all calculations the actual number of configurations is at least 10 000 and the first 5000 are discarded. The approach to equilibrium is illustrated by Fig. 2 taken from [1].

The thermodynamic properties may conveniently be obtained from the pair distribution function. This is defined by

$$g(r) = \frac{V^2 N(N-1)}{N^2 Z} \int \ldots \int \exp(-\beta U) \, dr_3 \, dr_4 \ldots dr_N,$$

$$r = |r_1 - r_2| \qquad (18)$$

where the integrations are carried out over all coordinates other than those of the interacting pair and U is given by the sum over pair potentials $\phi(r_i, r_j)$

$$U = \sum_{i<j-1} \phi(r_i, r_j) \qquad (19)$$

Z is the partition function, N is the number of particles in volume V. The factor $N(N - 1)$ is introduced because this is the number of pairs. In terms of the pair-distribution function $g(r)$ the average value of the potential energy is given by

$$\bar{U}/NkT = \frac{2\pi n}{kT} \int_0^\infty \phi(r)\, g(r)\, r^2\, dr. \tag{20}$$

The average value of the total energy in terms of kT is

$$\bar{E}/NkT = \tfrac{3}{2} + \bar{U}/NkT. \tag{21}$$

the pressure P is determined by

$$PV/NkT = 1 - \frac{2n}{3kT} \int_0^\infty r^3 g(r) \frac{d\phi(r)}{dr}\, dr \tag{22}$$

here n is the density. For a Coulomb system g must be replaced by $g - 1$ because of the effect of the background. The constant volume specific heat is given by

$$PV/NkT = 1 + \bar{U}/3NkT. \tag{23}$$

In the limiting case of the validity of the Debye–Hückel result, g_{DH} is given by

$$g_{DH}(r) = \exp[-(Ze)^2/4kT\varepsilon_0\, \mu\pi]\, \exp(-r/\lambda_D) \tag{24}$$

where the Debye length λ_D is given by

$$\lambda_D = (n_i Z^2 e^2/\varepsilon_0 kT)^{-1/2} = a(3\Gamma)^{-1/2}, \tag{25}$$

and Z is the ionic charge. In our units we may write

$$g_{DH}(x) = \exp(-\Gamma/x) \cdot \exp[-(3\Gamma)^{1/2}\, x] \tag{26}$$

$x = r/a$.

The agreement between the computations of Brush et al. and the Debye–Hückel result for $g(r)$ is shown in Fig. 3. The latter is in fact correct for $\Gamma \leqslant 0.1$. Results for various values of Γ obtained by B.S.T. are shown in Fig. 4 and are seen to differ radically for large Γ. Brush et al. also attempted to study solid–liquid transitions. It was shown by Hansen et al. [9] that more accurate computations are needed to get reliable results. Both the chain length and the particle numbers have been increased in more recent computations to obtain the small energy differences between the solid and liquid phases.

Starting with initial lattice configuration (body centred cubic) the system melted rather quickly for $\Gamma < 100$ and after 10^4–10^5 configurations a typical fluid type regime was found. At higher values of Γ the system melted either slowly or not at all [9]. Space forbids detailed discussion of the more recent

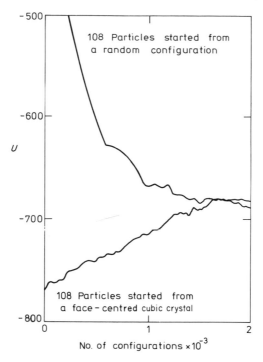

Fig. 2. Approach to equilibrium from a lattice and a random configuration $\Gamma = 10\cdot0$ for $N = 108$; U is the configurational energy of the system. (From [1], with permission.)

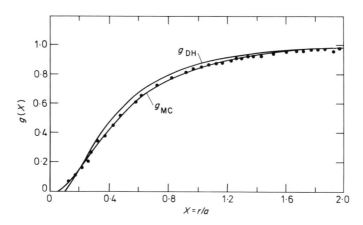

Fig. 3. Radial distribution functions $g(x)$ determined by the DH, and IDH equations and the MC result for $N = 32$ at $\Gamma = 0\cdot5$. (From [1], with permission.)

results but the equations computed by Hansen *et al.* for the equation of state will be quoted. The pressure is given by

$$P/NkT = 1 + U/NkT \qquad (27)$$

and the excess part U/NkT by

$$\frac{U}{NkT} = \Gamma^{3/2}\left(\frac{a_1}{(b_1 + \Gamma)^{1/2}} + \frac{a_2}{b_2 + \Gamma} + \frac{a_3}{(b_3 + \Gamma)^{3/2}} + \frac{a_4}{(b_4 + \Gamma)^2}\right). \qquad (27a)$$

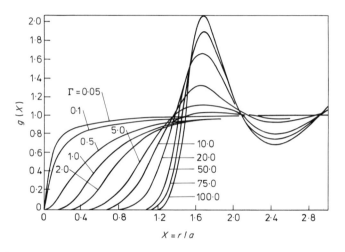

Fig. 4. Radial distribution functions $g(x)$ for $0.05 \leqslant \Gamma \leqslant 100.0$. (From [1], with permission.)

The coefficients can be found in [9]. More recently, DeWitt [4] has fitted Hansen's Monte Carlo data with the four-parameter function

$$U/NkT = a\Gamma + b\Gamma^s + c, \qquad (28)$$

where, for $1 \leqslant \Gamma \leqslant 40$, $s = 0.25$

$$U/NkT = -(0.89461 \pm 0.00003)\,\Gamma$$

$$+ (0.8165 \pm 0.0008)\,\Gamma^{1/4} - (0.5012 \pm 0.0016) \qquad (28a)$$

gives the best fit.

For $50 \leqslant \Gamma \leqslant 140$ the best fit was

$$U/NkT = -(0.8966 \pm 0.0001)\,\Gamma$$

$$+ (0.874 \pm 0.009)\,\Gamma^{1/4} - (0.568 \pm 0.023). \qquad (28b)$$

There seems to be a kink in Hansen's data at $\Gamma = 50$ which may indicate that 128 charges were not enough for large Γ. The heat capacity $C_v/Nk = d(U/Nk)/dT$ calculated from (28a) is given by

$$C_v/Nk = (1 - s)\,b\Gamma^s - c.$$
$$= 0\cdot6123\,\Gamma^{1/4} - 0\cdot5012. \tag{29}$$

There is, at present no clear theoretical model that can account for the $\Gamma^{1/4}$ dependence.

The ranges of validity of the model with a uniform electron background have been estimated in [1]. They are given in the form of a diagram (Fig. 5) of $\log kT$ (with kT in kV versus $\log n/n_0$ (with n_0 being the density at STP) for iron (^{56}Fe).

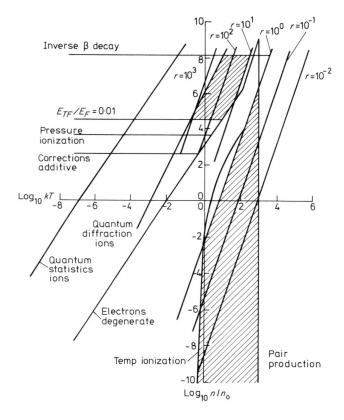

Fig. 5. Regions of validity of ion-background model for ^{56}Fe. The regions of validity are shaded. (kT is in kilovolts; n_c is the atomic density at STP). (From [1], with permission.)

Beyond 1000 kV, pair production becomes important. This range, although of astronomical interest, must be excluded. Above the pressure ionization line (when the Fermi energy of electrons is equal to the ionization potential) and above the degeneracy line for electrons they are considered to be uniformly distributed. Temperature ionization makes the distribution uniform when the Saha equation indicates 10% ionization. As far as quantum effects are concerned, quantum statistics for ions lies outside our range of interest. Quantum diffraction effects are considered important when the Einstein theory of specific heat indicates a 10% deviation from the classical value. For lighter elements, the pressure ionization line moves downwards and temperature ionization occurs more easily. The kink in the degeneracy line is due to relativistic effects. Clearly, in the region of interest, non-uniformity is important, and degeneracy effects have to be treated.

The Degenerate Electron Gas

The reader is supposed to be familiar with the elements of Fermi-Dirac statistics for electrons. A summary of results will nevertheless be given. A thermodynamic potential involving the distribution function (1) in the case of a continuous energy distribution is given by

$$\phi = \frac{8\pi V}{h^3} \int_0^\infty \ln\{1 + \exp[-(\beta p^2/2m - \alpha)]\}\, p^2\, dp. \tag{30}$$

This may also be written

$$\phi = \frac{V}{h^3}\left(\frac{2\pi m}{\beta\pi}\right)^{3/2} \chi(\alpha) \tag{31}$$

and with $\beta p^2 = 2tm$

$$\chi(\alpha) = \frac{4}{\sqrt{\pi}} \int_0^\infty \ln(1 + e^{\alpha - t})\sqrt{t}\, dt. \tag{32}$$

The number of particles is then

$$N = \frac{V}{h^3}(2\pi mkT)^{3/2} \frac{d\chi}{d\alpha} \tag{33}$$

and the energy

$$U = \tfrac{3}{2}NkT\chi(\alpha)\left/\frac{d\chi}{d\alpha}\right. \tag{34}$$

The equation of state is always

$$PV = \tfrac{2}{3}U. \tag{35}$$

The functions $\chi(\alpha)$ and $\chi' = d\chi(\alpha)/d\alpha$ have been computed by McDougall and Stoner [15] and Table II shows their dependence on α. Asymptotically

$$\chi(\alpha) \approx 2e^{\alpha}; \qquad \chi'(\alpha) \approx 2e^{\alpha} \quad \text{for} \quad \alpha < -4$$

$$\chi(\alpha) \approx \frac{16}{15\sqrt{\pi}}\alpha^{5/2}; \qquad \chi'(\alpha) \approx \frac{8}{3\sqrt{\pi}}\alpha^{3/2} \quad \text{for} \quad \alpha > 20.$$

Writing the distribution function in the form

$$f(\varepsilon) = \left[(\exp[\varepsilon - \xi(T)]/kT) + 1\right]^{-1} \tag{36}$$

with $\xi/kT = \alpha$, it is clear that in the limit of low temperature the particles fill all the lowest energy states ($f = 1$) and there are none above the Fermi

TABLE II. Values of the functions $\chi(\alpha)$ and $\chi'(\alpha)$.

α	$\chi(\alpha)$	$\chi'(\alpha)$
-4	0·036512	0·036397
-3	0·098713	0·097867
-2	0·26449	0·25860
-1	0·69335	0·65559
0	1·7344	1·5303
1	4·0045	3·1513
2	8·3308	5·6474
3	15·577	8·9751
4	26·521	13·023
5	41·829	17·688
6	62·077	22·893
7	87·776	28·581
8	119·39	34·710
9	157·33	41·248
10	202·01	48·169
11	253·79	55·453
12	313·03	63·080
13	380·06	71·037
14	455·21	79·309
15	538·78	87·885
16	631·08	96·754
17	732·38	105·91
18	842·98	115·33
19	963·14	125·03
20	1093·1	134·98

energy $\xi(0)$ which is given by

$$\xi(0) = \frac{h^2}{2m}\left(\frac{3}{8\pi}\frac{N}{V}\right)^{2/3}.$$
(37)

For higher temperatures, the degree of degeneracy may be measured by the ratio

$$\rho = \chi'(\alpha) = \frac{N}{V}\frac{h^3}{(2\pi mkT)^{3/2}}$$
(38)

which is $\frac{3}{4}\pi^{-5/2}$ times the number of particles in a sphere with radius equal to the De-Broglie wavelength.

A graph of $\ln \chi/\chi'$ taken from Sommerfeld [22] is displayed in Fig. 6.

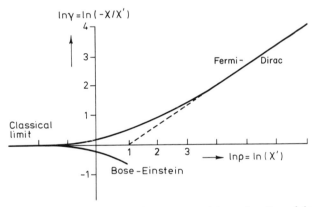

Fig. 6. Logarithm of the reduction factor $U/U_{\text{Boltzmann}} = \chi/\chi'$ as a function of the degeneracy parameter. (From [22], with permission.)

Since $\log U$ is proportional to the ordinate of this figure, and the abscissa to the degree of degeneracy it is seen that the deviation from the classical equation of state increases with ρ and $\rho(\partial U/\partial \rho)/U$ becomes constant in the limit of large ρ.

NON-UNIFORM ELECTRON GAS

We have seen above that there are important regions of parameter space where the electron distribution between the nuclei is not uniform. At high pressures, with lowered ionization energies and broadened spectral lines, when the distinction between core and valency electrons becomes unimportant, the electrons may be treated as a Fermi gas in the field of the nuclei.

This is the model of Thomas and Fermi which has been used in many situations. Surveys of such work were made by March [16] and by Brush [2]. Early pioneering work was done by Feynman *et al.* [6]; more recently Geiger *et al.* [7] published extensive calculations, their model is one of "quasi-atoms" formed by nuclei, surrounded by an electron cloud, with high probability found near the nucleus and shading of towards an atom radius. These quasi-atoms then undergo disordered motion due to the kinetic energy of the nuclei, but fit together continuously.

In the electron distribution (36) the energy now consists of potential and kinetic energy

$$\varepsilon = P^2/2m - e\psi(r) \tag{39}$$

and $\psi(r)$ is determined by Poisson's equation

$$\nabla^2\psi = \frac{1}{r^2}\frac{d}{dr}\left(r^2\frac{d\psi}{dr}\right) = 4\pi e n_E. \tag{40}$$

The electron density $n_E = N/V$ is given by (33) with α replaced by

$$\sigma = \alpha + \frac{e}{kT}\psi(r) \tag{41}$$

and the energy density is given by the modification

$$U_E = \tfrac{3}{2}kT\left[\frac{(2\pi mkT)}{h^3}\right]^{3/2}\chi(\sigma) - en_E\psi(r) \tag{42}$$

of (34). The equation for the potential distribution on the quasi-atom is then

$$\frac{1}{r^2}\frac{d}{dr}\left(r^2\frac{d\psi}{dr}\right) = 4\pi e\left[\frac{(2\pi mkT)}{h^3}\right]^{3/2}\chi'(\sigma). \tag{43}$$

The potential ψ is the sum of an electronic part ψ_E and a nuclear one $\psi_N = Ze/r$. Close to the nucleus, at $r = 0$, the nuclear part dominates and we have

$$\lim_{r \to 0}\{r\psi(r)\} = Ze. \tag{44}$$

On the other hand at the outer boundary $r = r_0$ electric field and potential vanish and we have

$$\psi'(r_0) = 0, \qquad \psi(r_0) = 0. \tag{45}$$

The differential equation (43) can be integrated numerically. It may also be written non-dimensionally

$$\frac{1}{\zeta^2}\frac{d}{d\zeta}\left(\zeta^2\frac{d\sigma}{d\zeta}\right) = \chi'(\sigma) \tag{46}$$

with $\xi = r/\lambda$, $\lambda = 1/2\pi e(h^6/8\pi m^3 kT)^{1/4}$. Numerically, it turns out that

$$\lambda = 1\cdot4044 \cdot 10^{-7}\, T^{-1/4},$$

and the boundary condition for $\xi \to 0$ is

$$\kappa = \lim_{\xi \to 0}\{\xi\sigma(\xi)\} = 1\cdot1903 \cdot 10^4\, \frac{Z}{T^{3/4}}. \tag{47}$$

Solutions of (46) with κ as a parameter have the form shown in Fig. 7 and the "quasi-atom-radius" is determined at the point $\xi = \xi_0$ where σ has a horizontal tangent. To illustrate results, we reproduce the radial electron

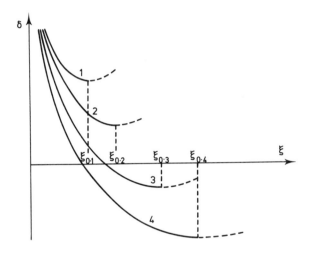

Fig. 7. Normalized potential σ versus normalized radius ξ for various values of κ. At the "quasiatom radius" the curves have horizontal tangents. (From [7], with permission.)

density for argon at different temperatures for a density of $1\cdot6 \cdot 10^{23}$ cm^{-3} in Fig. 8. The pressure in the dense plasma is related to temperature by the equation of state

$$p = \left\{\frac{\xi_0^3}{3}\frac{Z}{k}\chi(\alpha) + 1\right\}\bar{n}kT \tag{48}$$

derived from a simple picture of quasiatoms transferring momentum to an outer wall which is calculated from the momentum distribution of electrons at the quasi-atom radius. Other thermodynamic quantities can also be calculated and the monograph by Geiger *et al.* contains extensive tables. In order to understand the results of this theory more fully, it is instructive to discuss the equation of state (48). In the the case of strong degeneracy, with $\alpha \gg 1$,

$\sigma \gg 1$, the electron kinetic energy becomes proportional to $n^{2/3}$ and the potential energy proportional to $n^{1/3}$. Thus the fraction of potential energy decreases, the electrostatic field loses influence. The electron distribution becomes uniform and

$$n_E \to Z\bar{n} \quad (\bar{n} \text{ mean density}). \tag{49}$$

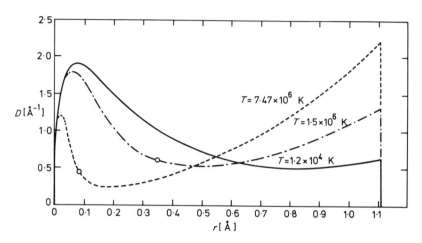

Fig. 8. Radial electron density for argon atoms for various temperatures. The number of particles in all cases is $n = 1.78 \cdot 10^{23}$ cm^{-3}. The points marked $\odot$ indicate the position of the classical radius. (From [7], with permission.)

The electron pressure P_E becomes proportional to the pressure P_N of the non-degenerate nuclei P_N, i.e.

$$P_E = \tfrac{2}{5} Z\alpha P_N = \frac{1}{20}\left(\frac{3}{\pi}\right)^{2/3}\frac{h^2}{m} Z^{5/3}\bar{n}^{5/3} \tag{49a}$$

and $P = P_E + P_N \to P_E$. In the case of a non-degenerate electron gas, for $\alpha = -4$, $\sigma = -4$, the electron density is again uniform $n_E = Z\bar{n}$, but now $P_E = ZP_N$. In between these two regimes we must resort to the results of computations, based on the work by Geiger et al. and equation (48) must be used to calculate the pressure. From the data of Geiger et al. [7] one can compute pressure as a function of log T and log n and we show results for Hydrogen and Neon in Figs 9 and 10. The regions of the diagrams of interest in connection with laser fusion are in a rectangle with $0 < \log kT < 5$ and $15 < \log N < 25$. The pressure is marked on the curves by a parameter given by P/nkT. Thus for Hydrogen, in the absence of degeneracy and non-uniformity effects, this parameter would be two for the sum of electron and ion pressures. It can be seen from Fig. 9 that, in the region of interest, the result varies between 1·5 and 5. In the case of Neon, purely classically, the parameter would be 11, whereas the computation shows it to vary between 3 and 40. It is now interesting to enquire whether the effect of plasma "structure", as given

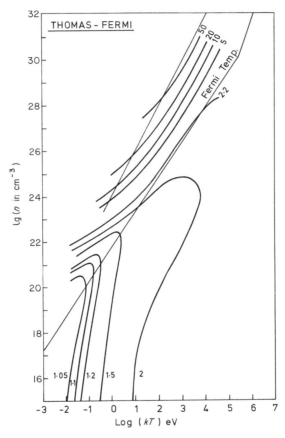

Fig. 9. Log n versus log T curves of constant p/nkT for hydrogen according to the Thomas–Fermi model.

by Hansen's Monte Carlo computations is marked. Using Hansen's formula (27), or DeWitt's formula (28) we find the plot given in Fig. 11 for hydrogen and in Fig. 12 for Neon. In these formulae, according to Hansen's suggestion, the pressure of a fully degenerate electron gas (49a) was added to that of the ions. This gives much lower values for the pressure than the Thomas Fermi calculation (Chapter 12, [4]).

It is interesting to note that the theoretical work discussed above leaves a considerable margin of uncertainty regarding the equation of state.

There are, however, further limitations of the TF model to consider. First of all the correlation effects: the electrons are not independent as they are

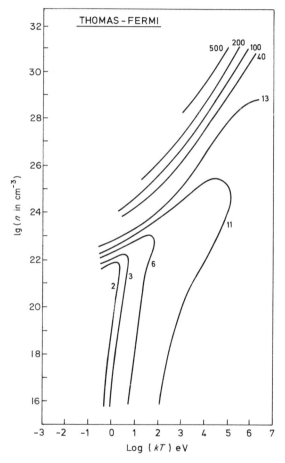

Fig. 10. Log n versus log T curves of constant p/nkT for neon according to the Thomas–Fermi model.

supposed to be in this model, but the position of one individual electron is dependent on the position of another so that a self-consistent description in which each electron is supposed to move in the field of all the others is incorrect. Secondly, the quantum mechanical effects have to be considered which arise when the wave nature of the electrons, which has been neglected in the TF semi-classical description, asserts itself. The correlation effects have two different causes, for one of the Pauli principle requires electrons with parallel spins to be kept at a greater distance apart than electrons in the singlet state. These are the exchange effects. The other causes are more dynamic in nature and arise from the coulomb interaction between electron pairs.

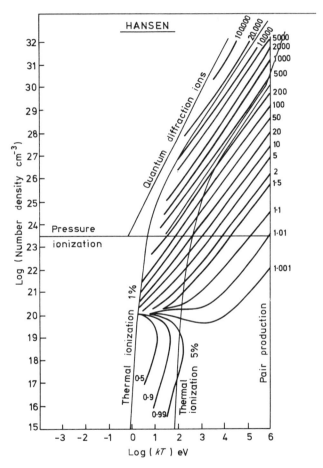

Fig. 11. Log n versus log T curves of constant p/nkT for hydrogen according to the Hansen model.

Dirac in 1930 suggested a method for improving the TF model to take account of exchange effects. The so-called Thomas-Fermi-Dirac model has been the basis of many computations. Quantitative results have been given by Latter [14]. Unfortunately when exchange effects arise quantum mechanical effects must also be expected and the accuracy of the corrections by this method are

in doubt. This is why the most detailed attention was given to the TF model
which has a further advantage that it scales simply with Z whereas in the
TFD model this property is lost. The region of applicability of the TF model
is shown as the part of shaded Fig. 13 [12]. In this figure the units were so

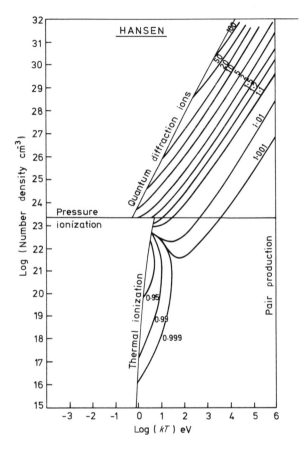

Fig. 12. Log n versus log T curves of constant p/nkT for neon rccording to the Hansen model.

chosen that $e = h = m = k = 1$. In order to express density in cm^{-3} and
temperature in K, the ordinate scale must be multiplied by $64/9\pi(e^2m^3/h^2) =$
$(2\pi)^3 \, 10^{18}$ and the abscissa by $4\pi^2e^4m/h^2k = 2{\cdot}09 \, . \, 10^6$.

Corrections to the TF model are discussed by Kirzhnik *et al.* [12] in
considerable detail.

Recently, More *et al.* [19] proposed a method for correcting the TF model
for the effects of ion movements. He considered that the ionic volume of the

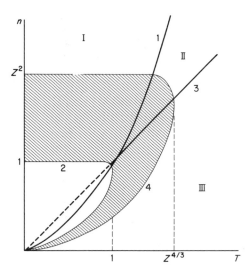

Fig. 13. Limits of applicability of the TFM. 1—degeneracy curve, 2—boundary of applicability of the TFM, 3—curve along which the exchange, quantum-mechanical and correlation effects are equal, 4—boundary of the uniformity regime. I—region in which exchange and quantum-mechanical effects dominate for a degenerate gas, II—the same for a Boltzmann gas, III—region in which correlation effects are dominant. The region in which the use of the TFM is justified is shaded.

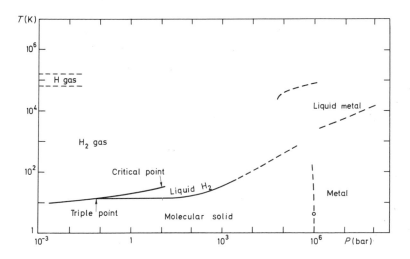

Fig. 14. Primitive phase diagram for hydrogen.

quasi-atoms considered by the TF theory is a quantity subject to fluctuations which he calculates according to a fluctuation formula to be found in Landau and Lifshitz. The probability of finding an ion occupying a volume V_a is found to be

$$\text{prob}(V_a) \approx \exp\{-[F_a(V_a, T) + PV_a]/kT\} \tag{50}$$

where P is the pressure of the fluid outside the ion. The most likely value of the atomic volume is given by the maximum probability requiring

$$P = -\left(\frac{\partial F}{\partial V_a}\right)_T. \tag{51}$$

In these formulae, F_a is the free energy and V_a is just the TF ion volume. More *et al.* average over the distribution (50) and find corrections which turn out to be small for pressures much in excess of the solid state pressure. Other authors have also computed corrections, e.g. Hansen [9] who merely corrected his classical Monte Carlo computations by means of an expansion taking account of quantum effects, due to Wigner [25] which is good when the De-Broglie Wavelength is small compared to the ion radius, or quasi-atom radius.

For the benefit of readers with a deeper interest in theoretical physics it may be noted that a complete quantum theory of electrons in a uniform ion background which contains all quantum and exchange effects exists. It is unfortunate that it has so far not been possible to deduce much quantitative information from the formal results obtained. One version of such a theory is due to Montroll and Ward and treated at length by Montroll [17]. The partition function is obtained from the solution of the so-called Bloch equation

$$HC = -\frac{\partial C}{\partial \beta} \tag{52}$$

for the generalized partition function

$$C(\mathbf{r}, \mathbf{r}', \beta) = \sum_i \psi_i^*(\mathbf{r}') e^{H\beta} \psi_i(\mathbf{r}), \qquad \beta = (kT)^{-1} \tag{53}$$

which reduces to the ordinary one for $\mathbf{r}' = \mathbf{r}$. This has the form of the time-dependent Schrödinger equation with β playing the role of time and the Feynman method for perturbation expansions as visualized by diagram methods is applicable. Recently Buckholtz [3], obtained semi-empirical results by means of similar expansions, and work relevant for the fusion track is being carried on by Graboske and DeWitt at Livermore.

In a paper by Salpeter and Zapolsky [21] the Thomas–Fermi method is generalized to obtain correlation corrections to equations of state at high pressure. Unfortunately, only low temperature results are given. The work is of importance for geophysical applications and a comparison with shock-wave data is made.

More accurate work on the effect of electron screening of nuclei in very dense plasmas has been done by several authors. Ross and Seale [20] used analytical perturbation methods. Much numerical and analytic work has been done by Hubbard and Slattery [10], Hubbard [10] and by DeWitt and Hubbard [27]. In the latter paper the metallic phase of light elements is discussed and computations are presented.

This brings us to a closely related subject, that of liquid metals. This is very well reviewed by Stishov [23] but there is also an older excellent review by March [16].

It is not possible to discuss here the relevance of this work, but a few remarks are in order. Along the fusion track, we must expect light elements to undergo phase changes. The transition to the metallic or semiconducting state of hydrogen and its isotopes may be accompanied by large changes in the thermal and electrical conductivity which are important for the dynamics of compression. Even helium may become metallic and in other light elements, outer electron shells will break open to set electrons free at sufficiently high pressure. These transitions have not been observed in the laboratory, and the whole subject of transport coefficients is a difficult one and will not be discussed here. It forms an important subject for future research. In the case of hydrogen, the metallic transition has been observed at low temperatures and at high pressures by Vereshchagin et al. [24], but the high-temperature behaviour is not known.

A phase diagram for hydrogen based on the extrapolation of existing data has recently been prepared by Leung et al. [26], Fig. 14, which shows the boundaries between metal, liquid metal, molecular solid, liquid and gas.

Yet another field of interest is that of the chemical changes at high temperatures and pressure. It may be possible to form compounds by laser compression which remain stable after rapid cooling, e.g. diamond is known to be formed at high pressure and the formation of hard carbides may be an interesting research topic.

REFERENCES

1. Brush, S. G., Sahlin, H. L. and Teller, E. (1966). *J. Chem. Phys.* **45**, 2102.
2. Brush, S. G. (1967). *In* "Progress in High Temperature Physics and Chemistry" (Cal. A. Rouse, ed). Pergamon Press, Oxford.
3. Buckholtz, T. J. (1971). UCRL-51055 Report, May.
4. DeWitt, H. E. (1976). *Phys. Rev.* A **14**, 1290.
5. Dirac, P. A. M. (1930). *Proc. Cambridge Phil. Soc.* **26**, 3, 76.
6. Feynman, R. P., Metropolis, N. and Teller, E. (1949). *Phys. Rev.* **75**, 1561.
7. Geiger, W., Hornberg, H. and Schramm, K. H. (1968). "Eng. der exakt. Naturwiss." (G. Höhler, ed.), Springer Tracts in Modern Physics. Springer, Berlin, Heidelberg.

8. Hammersley, J. M. and Handscomb, D. C. (1964). "Monte Carlo Methods". John Wiley & Sons, New York.
9. Hansen, J. P., Pollock, E. L., McDonald, I. R. and Vieillefosse, P. (1973). *Phys. Rev.* A **8**, 3096, 3110; (1975) *Phys. Rev.* A **11**, 1025.
10. Hubbard,W. B. and Slattery, W. L. (1971). *Ap. J.* **168**, 131.
    Hubbard, W. B. (1972). *Ap. J.* **176**, 525.
11. Hund, F. (1936). *Ergeb. Exakt. Naturw.* **15**, 189–228.
12. Kirzhnitz, D. A., Lozovik, Yu. E. and Shpatakovskaya, G. V. (1976). *Sov. Phys. USP*, **18**, 649.
13. Landau, L. G. and Lifshitz, E. M. (1958). "Statistical Physics", p. 255. Pergamon Press, Oxford.
14. Latter, R. (1964). *J. Chem. Phys.* **41**, 8, 2275.
15. McDougall, J. and Stoner, E. C. (1938). *Phil. Trans. Roy. Soc.* A **237**, 67.
16. March, N. H. (1957). *Advan. Phys.* **6**, 1.
17. Montroll, E. W. (1959). "La Theorie des Gaz Neutres et Ionisés". Hermann, Paris; John Wiley & Sons, New York.
18. Montroll, E. W. and Ward, J. C. (1957). *Phys. Fluids*, **1**, 55.
19. More, R. M. and Skupsky, S. (1976). *Phys. Rev.* A **14**, 474.
20. Ross, M. and Seale, D. (1974). *Phys. Rev.* A **9**, 396.
21. Salpeter, E. E. and Zapolsky, H. S. (1967). *Phys. Rev.* **158**, 876.
22. Sommerfeld, A. (1962). "Thermodynamik und Statistik". Akad. Verlagsges., Leipzig.
23. Stishov, S. M. (1975). *Sov. Phys. USP*, **17**, 625.
24. Vereshchagin, L. F., Yakovlev, E. N. and Timofeev, Yu. A. (1975). *JETP Lett.* **21**, 85.
25. Wigner, E. (1932). *Phys. Rev.* **40**, 749.
26. Leung, W. B., March, N. H. and Motz, H. (1976). *Phys. Lett.* A **56**, 425–426.
27. DeWitt, H. E. and Hubbard, W. B. (1976). *Astrophys. J.* **205**, 295.

12

COMPUTER CODES

The main purpose of this book is to give an insight into the physical processes relevant to laser fusion. Hence the emphasis is placed on analytic solutions but unfortunately, for practical purposes, numerical computations are necessary. One can, for the purpose of a better understanding, take the complicated mechanisms of the different processes apart and study them separately. Nonlinear processes can be dealt with, the influence of non-homogeneity can be studied, field effects, effects of fluid dynamic behaviour, etc., are well understood, but in the experimental situation all these components of behaviour occur simultaneously, in a way which makes the problem analytically intractable.

It should also be kept in mind, that the equations which are treated analytically are already based on approximations. Experiments in our field are rarely so clear-cut as to allow detailed checking of the theoretical assumptions. It is certain, however, that Newton's laws (or the mechanics of special relativity when velocities approach the velocity of light,) and Maxwell's equations apply. We can distinguish two types of computer codes, simulation codes, and others, by means of which solutions of differential equations are solved by means of finite difference schemes which are already approximate equations describing specific physical situations.

Simulation codes start from Newton's equations for particles which, by Maxwell's equations, are sources of fields when they carry charges; starting with an arbitrary particle and field configuration, orbits are determined computationally. The charge distribution is then obtained and from it, the field distribution. We have so far described the first iteration cycle. In the next, the orbits are determined in the fields just calculated, yielding a new charge and field distribution. This process is iterated and if it converges a solution is obtained which is much more free from assumptions than any solution obtained by solving more complicated differential equations which are themselves only approximately valid.

249

Unfortunately, in order to obtain meaningful solutions by the simulation process the number of particles, usually specified as the number within a sphere with a radius equal to the Debye length, must be large. The storage of the information regarding all the particles poses a severe problem and the computational effort is very large. One-dimensional codes are easier to handle, but many phenomena are essentially two- or even three-dimensional. So far codes are at best two-dimensional.

Passing to differential-equation codes, it must be stated at once that the space available in this book does not allow us to discuss numerical analysis. The reader will find a good introduction in the books by Potter [1] or by Fox and Mayers [2]. Differential operators like ∇^2 may be approximated in a finite difference mesh in which each point is surrounded by its neighbours as in Fig. 1, and the mesh length in our example is a (in two dimensions). It may be shown that, e.g.

$$\phi_P + \phi_Q + \phi_R + \phi_S - 4\phi_0 = a^2(\nabla^2\phi)_0 + \tfrac{1}{12}a^4(\nabla^4\phi)_0 + \dots.$$

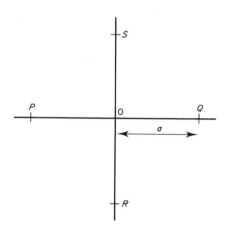

Fig. 1. Mesh points for computation.

Neglecting terms of higher order

$$(\phi_P + \phi_Q + \phi_R + \phi_S - 4\phi_0)/a^2 = (\nabla^2\phi)_0.$$

From the differential equations one obtains a set of difference equations whose solution converges to that of the differential equation for $a^2 \to 0$. The analysis of convergence is mathematically difficult. It is usual to regard convergence for successively smaller mesh length as a test of validity. There now exist well-tested standard routines for the solution of simultaneous linear and nonlinear equations.

We shall be content to state the equations which are solved by specific codes, in particular the Medusa code [3] which is well referenced and although it is only one-dimensional, it furnishes useful solutions to the compression and neutron-production problems. We shall also characterize a two-dimensional code.

THE PHYSICS OF THE MEDUSA CODE

The published Medusa code treats eight species: hydrogen, deuterium, tritium, ^{3}He (helium 3), ^{4}He (helium 4) and an arbitrary neutral atom N, of mass M_N and an arbitrary ion X with charge number Z_X and mass number M_X and neutrons (n). Species N and X play no role in the standard version but are carried along on the hydrodynamics in case one wishes to add impurities, e.g. for the purpose of spectroscopic diagnostics; they also allow the code to be run for elements other than hydrogen and its isotopes, omitting the nuclear reactions.

The instantaneous electron density n_e changes with time because of hydrodynamic expansion and contraction of the moving fluid elements. The local ion density n_i also changes due to atomic and nuclear reactions. The instantaneous chemical composition is described by a set of fractions f_k such that

$$n_k = f_k n_i \tag{1}$$

is the number density of ions of species k the fractions being adjusted to maintain the normalization

$$\sum f_k = 1. \tag{2}$$

The average mass and charge numbers

$$M = \sum_k f_k M_k, \qquad Z = \sum_k f_k Z_k \tag{3}$$

are used to compute the electron density $n_e = Z n_i$ and the physical density

$$\rho = n_i M m_H = \frac{1}{V} \quad \text{kg cm}^{-3} \tag{4}$$

where V is the specific volume. Electron-ion energy exchange, thermal conductivity, Bremsstrahlung, etc., also involve other averages such as $\langle M^p Z^q \rangle$.

The internal energy per unit mass (omitting subscripts i and e) is denoted by $U = pV/(\gamma - 1)\,\mathrm{J\,kg^{-1}}$ and the two subsystems, electrons and ions, have independent equations of state

$$U = U(\rho, T), \qquad p = p(\rho, T). \tag{5}$$

The energy equation is written as

$$C_v \frac{\mathrm{d}T}{\mathrm{d}t} + B_T \frac{\mathrm{d}\rho}{\mathrm{d}t} + p\frac{\mathrm{d}V}{\mathrm{d}t} = S \quad \mathrm{W\,kg^{-1}} \tag{6}$$

where S is the rate per unit at which energy enters each subsystem and

$$C_v = \left(\frac{\partial U}{\partial T}\right)_\rho, \qquad B_T = \left(\frac{\partial U}{\partial \rho}\right)_T. \tag{7}$$

The source terms S_e and S_i for electrons and ions are written respectively

$$S_i = H_i - K + Y_i + Q \quad \mathrm{W\,kg^{-1}} \tag{8}$$

$$S_e = H_e + K + Y_e + J + X \quad \mathrm{W\,kg^{-1}} \tag{9}$$

where H represents the flow of heat due to thermal conduction,

$$H = \frac{1}{\rho}\nabla\kappa\nabla T. \tag{10}$$

It has been pointed out by Salzmann ([8] of Chapter 8) that the electronic heat flux F_e can become unrealistically large when $\partial T_e/\partial r \gg T_e/\lambda_e$ where λ_e is the electron mean free path. The program limits the flux by setting it equal to F'_e given by

$$\frac{1}{F'_e} = \frac{1}{F_e} + \frac{1}{(F_e)_{\max}} \tag{11}$$

where $(F_e)_{\max}$ is given by

$$(F_e)_{\max} = \tfrac{1}{4}\beta_f n_e v_e kT. \tag{12}$$

According to elementary kinetic theory, one quarter of the electrons stream in one given direction, and the constant β_f is adjustable. Bickerton ([9] of Chapter 8) suggests $\beta_f = 0.02$ but higher values of β_f have been proposed by Haas et al. ([10] of Chapter 8) as being possible.

The program thus uses a modified heat conductivity

$$\kappa'_e = \kappa_e \left(1 + \beta_f \frac{\lambda_e}{T_e}\frac{\mathrm{d}T_e}{\mathrm{d}x}\right) \tag{13}$$

where

$$\lambda_e = 5 \cdot 7 \cdot 10^7 \ T_e^2/n_i Z^2 \quad \text{m}$$

$$\kappa_e = 1 \cdot 83 \cdot 10^{-10} T_e^{5/2} (\log \Lambda)^{-1} Z \langle Z^2 \rangle^{-1} \ \text{W} \text{m}^{-1} \ \text{K}^{-1}$$

$$\kappa_i = 7 \cdot 5 \cdot 10^{-12} T_i^{5/2} (\log \Lambda)^{-1} \langle M^{-1/2} Z^{-2} \rangle \langle Z^2 \rangle^{-1} \ \text{W} \text{m}^{-1} \ \text{K}^{-1}$$

where the heat conductivities are adapted from Spitzer ([4] of Chapter 2) to include suitable averages over the ionic charges.

The rate of absorption of laser light, X, is related to the laser power density $P_L(r, t)$ and the infinitesimal mass element dM by

$$X(r, t) = P_L(r, t)/dM \tag{14}$$

and $P(r, t)$ in turn to the power incident at the plasma boundary $r = R_0$ by

$$P_L(r, t) = \exp[-\alpha_0(R_0 - r)]P_L(R_0, t) \tag{15}$$

where α_0 is the absorption coefficient ((43) of Chapter 7). At the critical density surface $r = r_c$, all the remaining laser power is absorbed, i.e. $X(r_c, t) = P_L(r_c, t)$ to simulate anomalous absorption.

The terms Y_e and Y_i represent the rate of energy released from the four thermonuclear reactions (1)–(4) of Chapter 1. The number of reactions taking place is

$$R_{DT} = \langle \sigma v \rangle_{DT} f_D f_T n_i^2 \quad \text{m}^{-3} \ \text{s}^{-1}$$

with similar expressions for the other reactions. The neutrons are assumed to escape from the plasma carrying off their energy and this represents a mass and momentum loss. This is not quite correct because of scattering processes involving ions. (Compare Brueckner et al. ([12] of Chapter 1) for a more exact treatment). The charged reaction products are assumed to deposit their energy locally to the ions and electrons in a certain ratio. This again is not quite correct.

The chemical composition of the plasma changes due to these reactions and the program carries out the corresponding changes of the fractions f_T, f_D, etc. K is the rate of energy exchange between electrons and ions calculated according to a formula similar to (16) of Chapter 2, J the rate of *Bremsstrahlung* emission (see Chapter 1, equation (28)) and Q is the rate of viscous shock heating. The program uses

$$Q = b^2 \rho \left(\frac{\partial u}{\partial x}\right)^2 \frac{\partial V}{\partial t} \tag{16}$$

where Q is a viscous pressure and

$$\rho \frac{du}{dt} = -\nabla p \tag{17}$$

where u is the velocity in Lagrangian coordinates and b_i is a coefficient of viscosity. According to a well-known technique (see [1]) discontinuities due to shock waves are often avoided by the use of an artificial viscosity, a computational device simulating the effect of a physical viscosity which we have discussed in Chapter 9. (For the use of Lagrangian coordinates which follow the motion of fluid elements, compare the treatment of isentropic compression in Chapter 9). The pressure p is the sum $p = p_e + p_i$. The equation of state in the published program contains expansions of the Fermi–Dirac functions for the electron gas as used by Nuckolls et al. ([3] and (18) of Chapter 2).

In the program the dynamics are worked out separately for the two subsystems, electrons and ions, which are linked through their respective energy equations. Although the Medusa code is one-dimensional, it can also be used for compression with spherical symmetry, which requires a change in the coefficients of the differential operators, an option provided for in the program. It can, of course, be used for the solution of fluid dynamical computations, leaving out energy production.

Thus the results of analytic models for compression of spheres and shells can be checked. In spherical geometry, as has been shown in Chapter 9, adiabatic compression can be achieved by varying the pressure at the boundary with time (i.e. by providing a suitably shaped laser pulse) according to

$$p = 0_0\{1 - (T/T_0)^2\}^{-5/2} \tag{18}$$

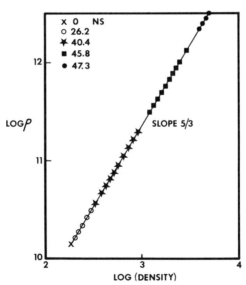

Fig. 2. Logarithm of pressure plotted gainst logarithm of density yields adiabatic exponent $\gamma = \frac{5}{3}$ with $s = -\frac{5}{4}$.

whereas in plane geometry, the appropriate boundary pressure is given by

$$p = p_0\{1 - T/T_0\}^{-5/4}. \tag{19}$$

Figure 2 shows the result of a computation for spherical geometry indicating that the compression is indeed adiabatic with the required slope $\frac{5}{3}$. The time of reaching positions on the curve is marked in ns.

Figures 3 and 4 show the result of testing the sensitivity of the result to a change in the exponent s of the boundary pressure law

$$p = p_0\{1 - (T/T_0)\}^{-s} \tag{20}$$

in the case of plane geometry. The curves deviate only slightly for changes up to $\pm 50\%$. This means that the theory is robust and that it should not be necessary to shape the laser pulse exactly in accordance with the ideal boundary pressure law.

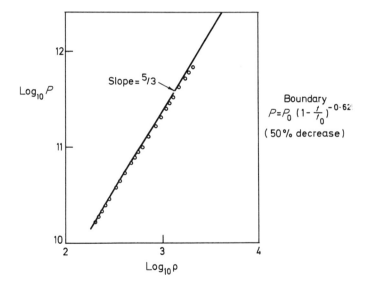

Fig. 3. Deviations from adiabatic of Fig. 2 with s increased by 50%.

Kidder ([14] of Chapter 9) has used his own code to calculate the time track of the particles in compression. The result, shown in Fig. 5 may be compared with the analytic result shown in Fig. 18 of Chapter 9 showing that the assumptions made in the theory were reasonable.

We now come to the discussion of computer results concerning compression and thermonuclear yield of fusion pellets. For the purpose of increasing the neutron yield, various types of pellets have been considered and research

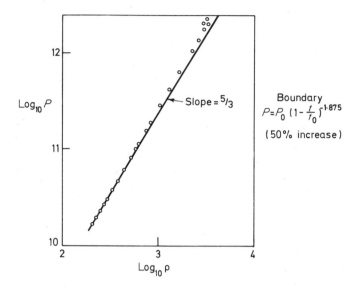

Fig. 4. Deviations from adiabatic of Fig. 2 with s decreased by 50%.

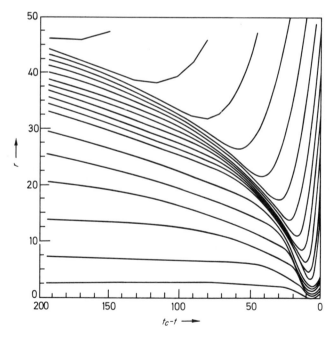

Fig. 5. Time tracks of imploding particles computed by Kidder ([**14**] of Chapter 9).

in this field is still progressing. In particular targets consisting of a succession of spherical shells of different chemical composition have been studied. Many experiments have been done with glass shells containing a mixture of deuterium and tritium gas. The outermost layer of solid or hollow targets may contain material of high atomic number and may act as a pusher to compress the fuel mixture inside. Modifications of the Medusa program can be used to study the compression of a variety of such targets.

To illustrate results we reproduce curves for the data of Brueckner and Jorna ([12] of Chapter 1) who give results obtained by means of their code. Bond *et al.* [4] have carried out the computations by means of the Medusa code. These data are given in Table I.

TABLE I

Solid DT sphere: initial radius 1·5 mm

Laser pulse form: two rectangular pulses and a ramp
$6\cdot3 \cdot 10^{11}$ W from 0 to 5·47 ns
$6\cdot3 \cdot 10^{12}$ W from 5·47 to 7·21 ns

Linear rise from
$6\cdot3 \cdot 10^{12}$ W at 7·21 ns to
$4 \cdot 10^{14}$ W at 7·42 ns

Initial temperature: 5000 K

The pulse sequence is used because a single shock cannot produce a compression ratio higher than 30 according to Goldman's result for $\gamma = \frac{5}{3}$ based on Gudderley's analysis [16] of Chapter 9. The final compression and neutron yield depend very much on the precise time sequence and the pulse shapes and optimization is carried out by varying the data inserted into the code.

The results of computations with the Medusa code are shown in Figs 6 and 7. The shocks move from right to left over distances indicated on the horizontal axis in metres. The numbers labelling the curves are the times in nanoseconds corresponding to the time steps after which computer results were printed out. Figure 6 shows the density profiles at successive instants and Fig. 7 completes the run with final time steps plotted with a contracted time scale. There is a sharp density increase which, by conduction etc, broadens at successive time steps. The lower curves of Fig. 7 show the result of a run for which the equation of state was changed. For ions the Hansen equation, (27) of Chapter 9, was used, while for electrons the Thomas Fermi data plotted in Fig. 9 of Chapter 9 were inserted into the program.

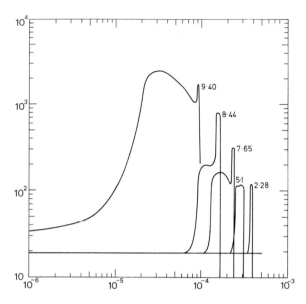

Fig. 6. Density ratio computed by Medusa code at successive times marked in ns on curves versus distance from centre in m (logarithmic scale).

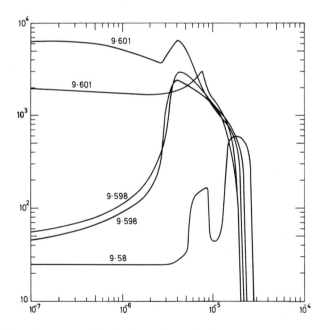

Fig. 7. Density ratio computed by Medusa code at the final stage with time marked in nanoseconds on curves. Tbe lower curves result when tbe equation of state is changed.

It will be seen that the final compression was somewhat reduced. However, this does not indicate that the optimized thermonuclear yield is necessarily reduced, because the output depends critically on the exact timing of the laser pulses. It is essential to ensure that the pulses arrive simultaneously at the centre. The following table illustrates this remark.

TABLE II

Starting time (ns)	Output power (kJ)
none	0·0002
9·10	0·044
9·19	1·6
9·250	229·00
9·26	11·7
9·27	2·9

It shows the energy output as a function of the starting time of the first shock when the second shock has been optimized with respect to the first. In this Lagrangian code the shocks are located by plotting constant mass points against time and the desired arrival time at the centre is obtained by a trial and error procedure.

Optimizing the pulses for the Hansen–Fermi–Thomas data an output of 322 kJ was obtained. The code results indicate that the temperature increases sharply from 10^7 to 10^9 K during the last few picoseconds. It will be seen that the maximum output power which we obtained by means of the Medusa code was 229·00 kJ. This contrasts sharply with the results published by Brueckner et al. ([12] of Chapter 1) of 510·0 kJ. These authors used a very different code which is not available and it is therefore not fruitful to speculate about the cause of this discrepancy. Superficially their curves look rather similar to ours, but our pulse shapes are sharper and propagate more slowly which may be due to our different scheme or to our treatment of viscous smoothing. It is very likely that their assumptions regarding energy deposition were different. The effect of changing the pellet radius was examined and the power output for pellet radii of 0·25 mm and 1 mm was remarkably low, as Table III shows.

In each case the optimization of the shock timing was carried out. We infer that Brueckner et al. (loc. cit.) had optimized their pulses for the radius 0·5 mm but that we had to optimize differently because our pulse propagation times were different.

It is clear that the physical assumptions entering any existing code are rather uncertain. We mentioned above that the constant β_f determining the

TABLE III

Pellet radius	Time of convergence (ns)	Power output (kJ)
0·25	4·771	16·0
0·50	9·601	229·0
1·00	19·367	45·1

flux limitation is not known. Ashby [5] concludes that ultra-high compression of DT is not possible for laser wavelengths of 1 μm or longer if Bickerton's value is correct. On the other hand, the role of the ponderomotive force which may well increase the efficiency of energy transfer to the core is not understood; the equation of state is only theoretically inferred, and conclusions regarding stability of the compression process (see Chapter 10) are only tentative.

Two-dimensional codes have been developed in many laboratories, but no fully published version appears to exist. A code called "Laser B" has been developed at Imperial College, London, [6] which has been used to study the effect of magnetic-field generation on laser interaction. We characterize it briefly.

DESCRIPTION OF THE PHYSICAL MODEL USED IN LASER B

In a two-dimensional domain (r, z) the following equations are solved,

$$\frac{\partial \rho}{\partial t} + \nabla \cdot \rho \mathbf{V} = 0$$

$$\frac{\partial}{\partial t}(\rho \mathbf{V}) + \nabla \cdot (\rho \mathbf{V}\mathbf{V}) + \nabla(P_e + P_i) = \mathbf{J} \times \mathbf{B}$$

$$\frac{\partial}{\partial t}\frac{P_i}{\gamma - 1} + P_i \nabla \cdot \mathbf{V} + \nabla \cdot \left[\frac{P_i \mathbf{V}}{\gamma - 1} + \mathbf{Q}_i\right] = \frac{n_e k(T_e - T_i)}{(\gamma - 1)\tau_{eq}}$$

$$\frac{\partial}{\partial t}\frac{P_e}{\gamma - 1} + P_e \nabla \cdot \mathbf{V}_e + \nabla \cdot \left[\frac{P_e \mathbf{V}_e}{\gamma - 1} + \mathbf{Q}_e\right] = -\frac{n_e k(T_e - T_i)}{(\gamma - 1)\tau_{eq}}$$

$$+ nJ^2 - P_{rad} + P_{las}$$

$$\frac{\partial B}{\partial t} = -(\nabla \times \mathbf{E})_\theta.$$

where

$$\text{electric field } \mathbf{E} = \eta \mathbf{J} - \mathbf{V}_e \times \mathbf{B} - \frac{1}{n_e e}\nabla(n_e k T_e).$$

The mass density is ρ, $\mathbf{V} = (V_r, 0, V_z)$ and $\mathbf{B} = (0, B_\theta, 0)$ in cylindrical coordinates. P_i and P_e are the ion and electron pressures, $\mathbf{Q}_i$ and $\mathbf{Q}_e$ are the ion and electron heat fluxes, P_{rad} is the *bremsstrahlung* radiation loss rate and P_{las} is the source of heating arising from the laser.

Some results are shown in Fig. 8. A laser beam with profile $I = f(t) \exp\left[-(r/r_b)^2\right]$ with $r_b = 20\,\mu m$ is incident on a thin foil initially positioned at $z = 40\,\mu m$. In a different program the light rays were traced by Snell's law as they go through the plasma. Figure 8 shows the plasma variables V_r, V_z, B and T_e at 75 ps just after a "hole" has been burned through the foil. Notice from Fig. 8(D) that the plasma is hot only where the laser beam is situated. This is due to the magnetic field with a configuration shown by Fig. 8(C). All the figures have rotational symmetry, and the magnetic field has an off-axis maximum. It acts to inhibit the heat flux; according to theory the reduction factor is given by $1/(1 + \omega_{ce}^2 \tau_{ei}^2)$. Figure 8(E) shows the electron density with the axial hole. Figure 8(F) shows bursts of fast ion flux as a function of time, coming from the front of the foil. There are also bursts from the rear. The axial velocity of the plasma (Fig. 8(B), is directed towards the incident laser beam at the front and away from it at the rear side.

RESULTS FROM SIMULATION CODES

The inverse *bremsstrahlung* absorption coefficient is too low to account for the results of measurements of absorption ([5] of Chapter 3) which range from 20–50% for Nd light. The theory of anomalous absorption processes has been discussed at length in Chapter 7. Some results from two-dimensional computer simulations have been reported by Kindel *et al.* [7], others by Valeo *et al.* [8] in a paper which contains many references to the literature on this subject. The light intensity assumed was $10^{16}\,W\,cm^{-2}$ ($\lambda = 1\cdot02\,\mu$) and the plasma slab had a linear density profile with scale length $L = 10\lambda/2\pi$. Results consistent with the expected nonlinearly steepened density profiles were obtained. The absorption peaks at the angle of incidence predicted by the theory outlined in Chapter 7 (see (75) of Chapter 7). It is estimated that about 25% net absorption is contributed by the various processes leading to plasma wave generation. The paper also contains estimates of hot electron generation by the instability processes. As has been explained already, simulation is similar to experiment, in that no assumptions about the physical processes are made and the results include their combined effects.

At the end of Chapter 6, attention was drawn to a paper by Estabrook which confirms conjectures about bubble formation by trapping of radiation, the action of the ponderomotive force, etc. This work was done by means of a simulation code described by Langdon and Lasinski in [9].

K

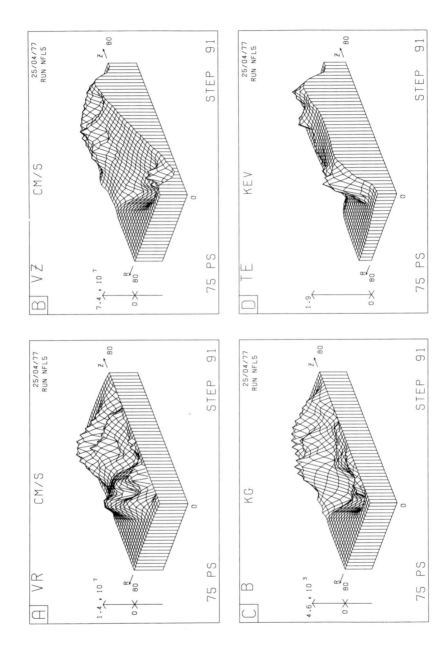

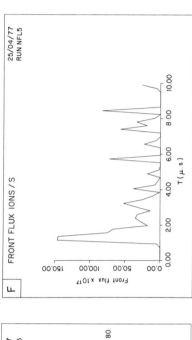

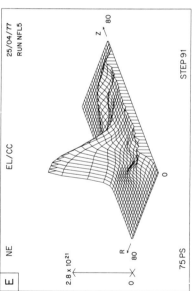

Fig. 8. Results from 2-dimensional Laser B code. A: radial velocity, B: axial velocity, C: magnetic field, D: temperature, E: electron density, F: bursts of ion flux.

With a density gradient of scale length greater than two wavelengths, the ponderomotive force due to the Airy pattern of the incident and reflected light and the Brillouin instability caused the ions to form large amplitude waves; eventually the critical density is exceeded, light is trapped which then drives a sausage type of instability so that the light pressure changes the density trough to density pockets or bubbles. The bubbles grow and eventually fuse to form bigger bubbles. More complicated phenomena and the limitations of the method are also discussed in the paper. It is not known what the subsequent development might be after the formation of bubbles or whether such entities are stable. Although two-dimensional simulations cannot give any definitive answers, computations by Forslund *et al.* reported by Lindman ([4] of Chapter 13) are interesting. Figure 9(a) shows the case when the electric field of the laser light is perpendicular to the propagation

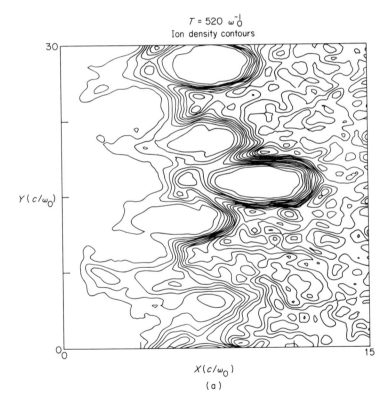

$$T = 520 \; \omega_0^{-1}$$
Ion density contours

Fig. 9(a). Density contours in x–y space from a two-dimensional point in cell simulation of the surface rippling instability. The light wave is normally incident and its electric field vector is in the z-direction. (From [4] of Chapter 13, with permission.)

$$T = 810 \ \omega_0^{-1}$$

Ion density contours

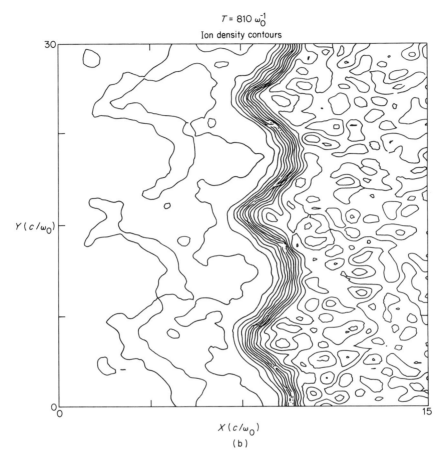

$X(c/\omega_0)$

(b)

Fig. 9(b). Density contours in x–y space from a two-dimensional point in cell simulation of the surface rippling instability. The light wave is normally incident and its electric field vector is in the x–y plane at an angle of 45° to the z-axis. Saturation at low amplitude is observed in this case. (From [4] of Chapter 13, with permission.)

vector **k** of the surface perturbation. Initial fluting is followed by bubble formation and general destruction of the plasma vacuum interface. When E is parallel to **k**, no instability is observed. When the electric field is at an angle of 45° to the **k** of the surface perturbation the surface is unstable. The perturbations shown in Fig. 9(b) saturated at low amplitude.

REFERENCES

1. Potter, D. E. (1973). "Computational Physics". Wiley–Interscience, London.
2. Fox, L. and Mayers, D. F. (1968). "Computational Methods for Scientists and Engineers". Oxford University Press, London.
3. Christiansen, J. P., Ashby, D. E. T. F. and Roberts, K. V. (1974). *Computer Phys. Comm.* **7**, 271–87.
4. Bond, R. and Motz, H. (1978). *J. Phys.* D, **11**, 1795–1807.
5. Ashby, D. E. T. F. (1976). *Nucl. Fus.* 623.
6. Private communication from Dr J. Kilkenny Imperial College, London.
7. Kindel, L. M., Lee, K. and Lindman, E. L. (1975). *Phys. Rev. Lett.*
8. Estabrook, K. F., Valeo, E. J. and Kruer, W. L. (1951). *Phys. Fluids*, **18**, 1151–1159.
9. Langdon, B. F. and Lasinski, A. B. (1975). *Bull. Amer. Phys. Soc.* **20**, 1377.

13

RESULTS FROM EXPERIMENTS

In this chapter the salient features of the results from experiments will be reviewed. Work in this field is in rapid process and the conclusions must therefore be tentative. It is not possible within the space available to give details of all the experimental methods not to do justice to the numerous authors. At any rate, a comprehensive review would be premature at the present stage. Material will be used up to the time of the Oxford Conference in 1977.

Most of the work so far has been done with plane targets or with spherical targets in the exploding pusher mode. There is evidence for the temperature of the plasma in the interaction region, mostly from X-ray spectra and electron spectra. The interpretation of the observation of high energy electron emission yields information regarding the light absorption process. Further evidence is provided by the study of the angular dependence and polarization of the reflected laser light. Fast ions emitted during the implosion of targets are an important item in the energy balance. Some information on the time evolution of the plasma can be obtained from the time resolved observation of ion spectra. X-rays emitted by the exploding and imploding portions of the exploding pusher targets can be imaged by means of a simple pinhole camera. Images at different times can be obtained by means of streak cameras [8] to show the time evolution of the process. The X-rays from various spacial locations can also be spectrally analysed. The neutron yield has been measured and the temperature of the dense core can be inferred by comparison with results from codes and from neutron energy distribution measurements. Alpha particles have been imaged by means of Fresnel zone plates and their thermal spread has been measured and used for core temperature determination.

Temperature Determination by X-ray Cameras

Most experiments were done with a laser flux of 10^{15}–10^{16} W cm^{-2}. Plasma temperatures in the corona have been found to lie between 400 eV and 1 keV. From the wealth of experimental data, we quote some results from the Rutherford Laboratory. In Chapter 2 we illustrated the target chamber (Fig. 1)

DOUBLE BEAM FOCUSSING

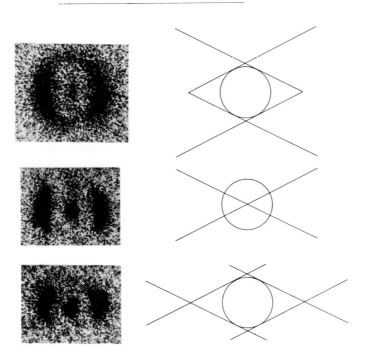

Fig. 1. On the left: X-ray pinhole camera pictures of an imploding hollow sphere target; on the right: schematic illustration of the corresponding focusing arrangement. Topmost: double focusing behind the target; middle: Focusing on the target centre; bottom: focusing in front of the target (Rutherford Laboratory.)

where two lenses illuminating the target can be seen. Glass microballoon targets typically of 70–100 μm diameter and 1 μm wall thickness were used, which were mounted on a carbon fibre stalk. They could be filled with D_2, $D_2 + T_2$, neon or argon. The two beams are focused by $f/1$ lenses and they are aligned to produce a focus in front, behind or in the middle of the target.

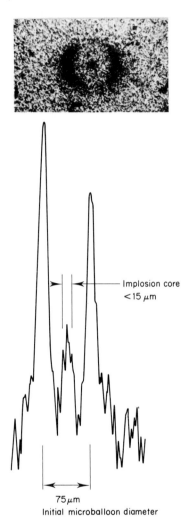

Fig. 2. On top: X-ray pinhole camera picture of an imploding 75 μm diameter hollow sphere (glass microballoon). Below: densitometer trace showing peaks due to the glass shell and the compressed core (15 μm) (Rutherford Laboratory.)

X-rays from the target are imaged by means of a pinhole camera and provide a picture of the result of the explosion. Figure 1 shows the pictures obtained with the three types of focusing. The innermost spot is produced by X-rays from the collapsed core and the outer shells are images of the exploding glass, and it is seen that images behind the target produced the most symmetrical explosion. Figure 2 shows a densitometer trace of the photograph.

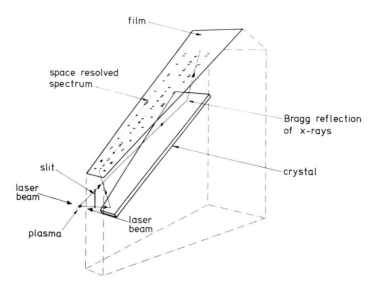

Fig. 3. Miniature space-resolving X-ray spectrometer. The X-rays from the laser irradiated plasma are Bragg-reflected from a crystal and photographed on a film (Rutherford Laboratory).

The apparatus used for space resolution of the picture is schematically shown in Fig. 3. A thin slit is used for lateral imaging. The figure shows the photographic film situated above a crystal from which the X-rays are reflected under the Bragg angles corresponding to the various wavelengths. The resulting record is schematically illustrated by Fig. 4.

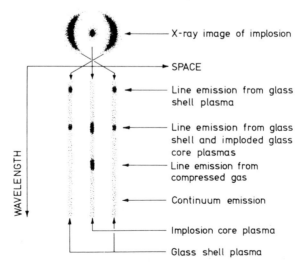

Fig. 4. Schematic of space resolved spectrum (Rutherford Laboratory).

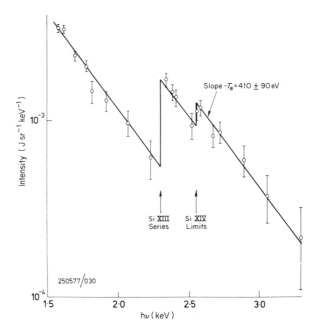

Fig. 5. Space-resolved X-ray continuum spectrum of implosion core of glass microballon with 8·5 bar neon fill (Rutherford Laboratory).

Figure 5 shows a space-resolved X-ray continuum spectrum, showing absorption edges and recombination radiation continua from highly stripped Si ions. The instrument was absolutely calibrated and since the oscillator strength of hydrogen-like spectra are known, the density in the source region can also be determined since its volume is also known from pinhole camera measurements. The temperature is determined from the slope of the recombination spectrum shown in Fig. 5. Line width of the Si XIII

TABLE I. Glass Shell Plasma—Shot No 250577/30.

Te	$700 \pm 200 \, \text{eV}$	from Si recombination continuum slope
Te	$730 \pm 30 \, \text{eV}$	from Si XIV Ly γ. Si XIII 1 $^1S_0 - 4\,^1P_1$ intensity ratio
$N_e^2 V$	$1\cdot4 \pm 0\cdot4 \times 10^{37} \, \text{cm}^{-3}$	from absolute intensity Si XIII recombination continuum
V	$1\cdot5 \times 10^{-8} \, \text{cm}^{-3}$	from pinhole camera images and space resolved spectra
N_e	3×10^{22} electrons cm^{-3}	from above

$1^1S_0 - 3^1P_1$ emission and intensity ratio of the Si XIV Lyman: Si XIII $L^1S_0 - 4^1P_1$ was also used for temperature determination. Typical results are shown in Table I.

Monitoring a very large range of X-ray energies with a much simpler apparatus using absorbing metal foils yields plots of relative X-ray intensity versus energy as determined by the cut-off energies of foils of various thickness. Such a plot is shown in Fig. 6 for a laser power of 10^{15} W cm^2. A region of big slope corresponds to the low temperature measured more accurately by the spectroscopic method described above. From the low slope region a temperature of 13·5 keV is determined and this must be due to the fast electron component. It originates from the regions of high ion density in the core and outer shell.

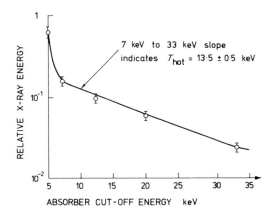

Fig. 6. Relative X-ray energy versus the cut-off energy in eV of the absorber foils (Rutherford Laboratory).

Electron spectra have been observed, e.g. by Giovanielli, *et al.* [1] at Los Alamos and Giovanielli [2] has gathered data from many laboratories, some obtained with CO_2 lasers and others with N_d lasers, and displayed them in a single graph shown in Fig. 7. The logarithm of the hot electron "temperature" is plotted against $\phi\lambda^2$ where λ is the laser wavelength, and this quantity is proportional to the energy of electrons oscillating in the electric field of the laser. Results obtained with the two types of laser then lie on the same graph.

The fact that the hot electron temperature can be expressed in the form

$$T_{\text{hot}} \propto (\phi\lambda^2)^\delta$$

with δ taking values 0·68 and 0·25 in different regions of the plot has been interpreted in Chapter 8 to mean that the portion with $\delta = 0.67$ represents a regime where the laser flux is balanced by free streaming electron heat conduction while for higher values of the ponderomotive force steepened the profile and resonance absorption becomes prominent.

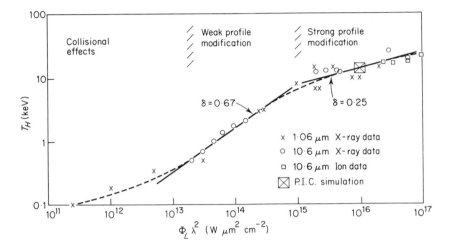

Fig. 7. Hot electron energy T_h in keV plotted against the product of laser flux and squared laser wavelength in W μm cm^{-2}. The curve is compiled from results obtained in many laboratories.

ION SPECTRA

Ion spectra have been measured both in plans target and in spherical compression experiments. Recent work by Rumsby *et al.* at the Rutherford Laboratory [3] was done with a Nd laser by means of Faraday cups for collection of ion charge coming from the target in various spacial directions. Electrons were excluded by the experimental technique. Figure 8 shows signals from these collectors for the four directions indicated correspond to those of the east and west beam of the laboratory. The current pulses recorded show a reproducible three-component structure. Velocities at the current peaks are 2.10^8, 5.10^7 and 10^7 cm s^{-1} respectively corresponding to energies for oxygen ions of 320, 20 and 0·8 keV. These three components may be assumed to originate in hot electron plasma created by the absorption process, in the ablating thermal plasma and in the imploded core respectively. The first two components are also observed with plane targets, but the third one is a feature of the implosion events only. It is difficult to base conclusions about the nature of the generation mechanism on current collected by distant probes, because of acceleration experienced by the ions during their flight away from the target. For highly stripped ions the M/Z ratio is 2. The data have been analysed on this assumption to give mass and energy flux rates. In this case the fast ion component typically carried 60% or more of the total energy recovered by the probe but has only a few per cent of the mass. Assuming that the fast ions were protons would reduce the percentage to 40.

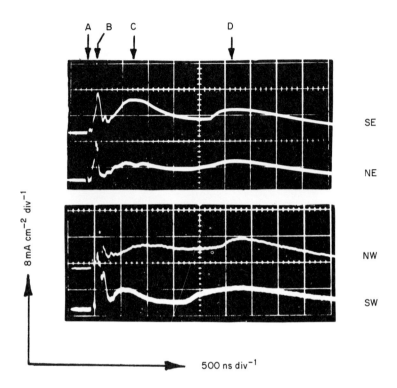

Fig. 8. Signals from ion collectors. The two laser beams are incident from the laboratory east and west direction. Ions collected from the directions indicated. A: laser time marker, B: fast ion signal, C: normal ion signal, D: slow ion signal; $2 . 10^8$, $5 . 10^7$, 10^7 cm s^{-1} respectively (Rutherford Laboratory).

Ion collection is an important technique for determining the total absorption of laser light by the target. Most of this adsorbed energy (90%) eventually reappears in the form of ion energy and about 1% is turned into X-ray energy.

Data on fast ion generation by a CO_2 laser on a plane CH_2 target published by Lindman [4], are produced in Fig. 9. The solid line is number density of ions as a function of velocity calculated by means of the free streaming solution of Chapter 8.

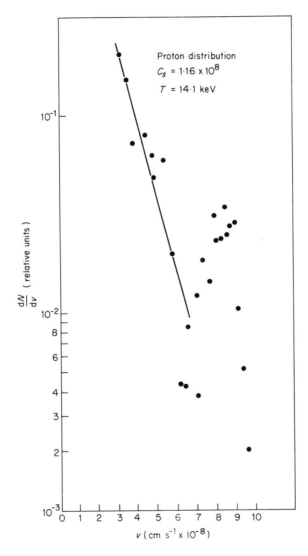

Fig. 9. Number of hydrogen ions as a function of ion velocity from a plane CH_2 target at an intensity 10^{15} W cm^{-2}. Straight line as a fit using a self similar solution. Deviations from straight line may be due to breakdown of quasineutrality [4].

ABSORPTION

Evidence for resonance absorption from plane target experiments with a Nd
laser has been reviewed by Ahlstrom [5]. The strongest evidence for resonance
absorption is furnished by results on the angular dependence of the absorption
of polarized laser light shown in Figs 10 and 11 respectively.

They show the results of measurements at 1·06 μm (Nd Laser) by Balmer
and Donaldson [9] carried out with a laser energy flux of 2.10^{13} W cm^{-2}. A
plane target could be rotated to vary the angle of incidence. The plane of
polarization of the light beam could be rotated, so that measurements could
be made for the case of s-polarization (E-vector perpendicular to the plane of
incidence) and for p-polarization (E-vector parallel to the plane of incidence).

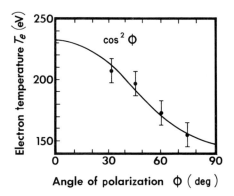

Fig. 10. Dependence of X-ray energy (i.e. plasma temperature) on the angle ϕ between E-vector
and plane of incidence. (From [9], with permission.)

The energy of the X-rays emitted by the foil was measured by absorber foil
techniques as a function of the angle of incidence and also as a function of the
angle of polarization. The X-ray energy is a measure of the plasma temperature
and thus indicates the degree of laser light absorption.

The results of such measurements at a fixed angle of incidence θ (15°) are
shown in Fig. 10 as a function of the angle of polarization ϕ. The cos² ϕ
curve represented by the continuous line is seen to be a good fit to the data.
This demonstrates the presence of resonance absorption which should vary
as the square of the E-vector component in the plane of incidence given by
$E_0^2 \cos^2 \phi$. Figure 11 shows the results for p- and s-polarization as a function
of the angle of incidence θ. It is seen that the dependence on angle θ shows a
peak between 15° ± 5° when the E-vector is in the plane of incidence (p-
polarization), but is monotonic in the case of polarization perpendicular to

this plane (s-polarization). The theory of resonance absorption of Chapter 7 leads to the angle

$$\theta = \arcsin \frac{0 \cdot 8}{(Kr_0)^{1/3}}$$

of maximum absorption where r_0 is the scale length of the inhomogeneity and it is interesting that the measurement thus allows to determine it to be $\approx 3\,\mu m$.

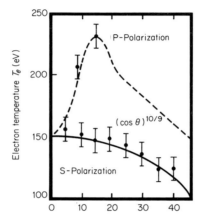

Fig. 11. Dependence of X-ray energy, i.e. plasma temperature on the angle of incidence θ for s- and p-polarization. (From [9], with permission.)

Evidence for the occurrence of Brillouin scattering is provided by the wavelength shift (due to the emission of an ion sound wave) which has been seen in some experiments. Experiments with the Cyclops laser on glass with polarization yielded polar scattered light distributions which are elliptical with the major axis perpendicular to the plane of polarization. The large ratio 4:1 of major and minor axis is indicative of Brillouin scattering. Indeed, for coupling with the backscattered wave to occur, its electric field vector must have a component in the direction of that of the incident field and this is the case for p- but not for s-polarization for 90° scattering. For s-polarization, no enhancement of 90° scattering is observed.

Long pulses and large spot size would tend to spread the plasma out more evenly and lower the Brillouin threshold compared with that obtaining for steep density gradients. Experiments on glass spheres with Argus shows a degrading of absorption, i.e. increased reflection under such circumstances thus giving further evidence for Brillouin scattering. On the whole experiments are not yet conclusive.

PROFILE STEEPENING

Some evidence concerning profile steepening comes from the study of the polarization of the reflected laser light. In the extreme case of an infinite density jump as in the case of reflection from a metal surface components of the field must vanish and the field vector of the reflected light is rotated by 180° with respect to that of the incident light. On the other hand, the electric field vector of light obliquely incident on a medium with a slowly varying index of refraction maintains its angle with respect to the scattering plane as it refracts uniformly. Thus only the component parallel to the target surface is reversed. Phillion et al. [10] have examined this problem and obtained evidence for steep gradients, by means of experiments performed on Argus. The best evidence is however obtained by means of holographic studies. A diagnostic laser beam is made to pass near the target during the pulse and it is recorded together with a reference beam. From the resulting interference fringes the profile can be reconstructed. (In the case of rotational symmetry the method relies on a mathematical theorem due to Abel [7] and is therefore called Abel-inversion). Attwood et al. [12] report a scale length of $1 \cdot 5 \, \mu m$ seen in this way and as some evidence for rippling of the critical density surface. A 45-m glass shell was irradiated by means of the Janus laser system with approximately $10^{14} \, W \, cm^{-2}$. A portion of the main Nd beam was split off and the 30 μs pulse was passed through two successive frequency doubling

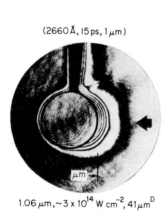

(2660 Å, 15 ps, 1 μm)

$1.06 \, \mu m, \sim 3 \times 10^{14} \, W \, cm^{-2}, 41 \mu m^{D}$

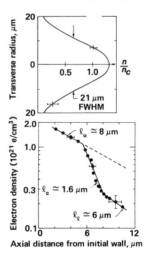

Fig. 12. Profile steepening due to radiation pressure. $3 \cdot 10^{14} \, W \, cm^{-2}$ of Nd laser light incident on 41 μM diameter microballoon. On the left: Interferogram. On the right, top figure: electron density in units of the critical density as a function of transverse distance. Bottom figure: electron density as a function of axial distance from initial wall showing profile steepening. (From [12], with permission.)

crystals and a frequency quadrupled probing pulse was obtained with 1 % conversion efficiency.

This makes the frequency four times the critical frequency ω_c and with the probing pulse shortened to 15 ps an interferogram was obtained which, when inverted by the Abel inversion method showed a steepened region between $1\cdot1\ n_c$ and $0\cdot3\ n_c$ with a scale length of $1\cdot6\ \mu$m. A typical result of interferometry is shown in Fig. 12.

Most experiments to date have been done in the explosive pusher mode, because this is the easiest way to achieve symmetry. The details of some of the targets used in the experiments at Livermore have not been published but the achievements have been summarized by Ahlstrom [5] in a contour map (Fig. 13) the $n\tau$ produce against ion energy. Under the name of the facility the code names of the targets have been written. The parameter labelling the curves is the gain factor, which is the ratio of the neutron yield to the energy delivered to the target. Also included in the map are the results to be hoped for in the future when the Shiva system is commissioned.

LASER FUSION – PROGRESS, PROJECTIONS

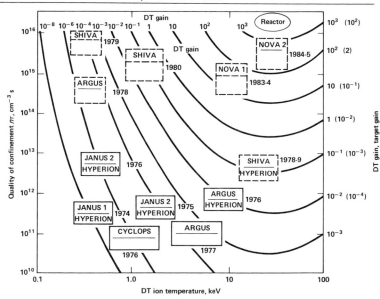

Fig. 13. $n\tau$ product obtainable from present and future laser facilities and targets versus DT ion temperature [5].

The best results so far are temperatures of 10 keV and a gain of 1 %. It is seen that a gain of unity or scientific break even is projected for a date not too far in the future at the time of writing but there is still a long way to go until a gain of 10^3 can be achieved.

Ahlstrom also presents a graph Fig. 14, in which a normalized neutron yield against the useful specific energy, ε_c, that is the energy absorbed per unit of target mass. It is seen that the neutron yield scales with the initial radius r to the 10/3 rd power and with wall thickness w to the 3/2 th power. It is given by

$$N = 9 \cdot 6 \cdot 10^6 \, r^{10/3} w^{2/3} \varepsilon_c^{-7/6} \exp(-5 \cdot 45/\varepsilon_c^{1/3})$$

where r and w are in μm, and ε_c in J ng^{-1}.

NEUTRON YIELD VS USEFUL SPECIFIC ENERGY

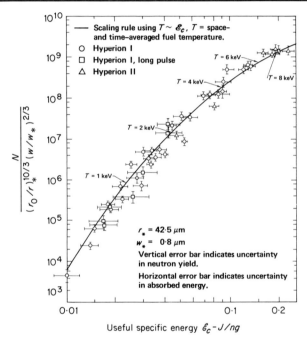

Fig. 14. Neutron yield versus useful specific energy [5].

These data summarize the present position. It should be borne in mind that pressure density curves achieved with the explosive pusher type of implosion are far from what must be aimed at. Ahlstrom gives an interesting survey of the p–n curves achieved in the Livermore experiments (Fig. 15) which also shows that adiabat projected for future devices in the planning stage, Shiva and Nova, which approach the adiabat of a Fermi-degenerate electron gas. The discussions of Chapters 11 and 12 show that the equation of state for DT is not likely to differ from this in a practically significant way.

So far we have concentrated on results and future plans of the Livermore Nd glass laser facilities. It has already been pointed out that glass lasers cannot reach the efficiency needed to obtain net reactor gain. This might be possible with gas lasers, e.g. the CO_2 laser. Profile-steepening observed in many laboratories has led to the hope that such lasers will couple as well to the

FUEL ADIABATS OF LASER FUSION TARGET DESIGNS

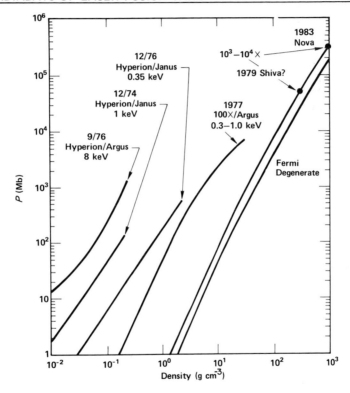

Fig. 15. Full adiabats of laser fusion target designs [5].

compression target as glass lasers. Experiments have been carried out at Los Alamos which have strengthened confidence in this respect to such an extent that very large CO_2 installations have been built and one, Antares, designed for operation in the 100 terawatt region is in the planning stage. Figure 16 shows the development schedule at Los Alamos and the kind of experiments which are planned for the various stages are indicated [11].

A very promising candidate is the Free-electron Laser. This is a purely electromagnetic device, in which a high energy electron beam is made to carry out a sinusoidal or spiral motion by means of a periodic array of magnets. It was shown in Chapter 5 that such a beam radiates at a frequency

$$\omega = (2\pi\bar{v}_z/L)(1 - \bar{v}_z/c)^{-1} \approx 2\bar{v}_z(2\pi/L)(1 - \bar{v}_z^2/c^2)^{-1} = 2\bar{v}_z(2\pi/L)\gamma^2$$

where L is the spatial period of the orbit. It may also be shown, by an extension of the theory presented on pp. 68–70 of that chapter, that it will amplify laser radiation. In the case of transverse waves the spontaneous transition probability becomes

$$w_p^{\delta} = \mu_0\pi^2 c\bar{v}_z^2 e^2 A^2/(\hbar kL^2)\delta(\omega - \omega_0 - k\bar{v}_z)$$

where A is the amplitude of the sinusoidal motion, where $\omega_0 = 2\pi\bar{v}_z/L$ and $\bar{v}_z$ is the mean axial velocity of the electrons. The growth rate is obtained by substituting $w_p(k)$ into Formula (48a) of Chapter 5. The details may be found in [14]. The amplification of light was experimentally demonstrated by Madey et al. [15] using a 40 MeV beam to amplify light from a CO_2 laser ($\lambda = 10.6\,\mu m$). They also generated infrared radiation at $3.4\,\mu m$ at a high power level. Although the radiation energy emitted is only a small fraction of the beam energy, the device may be made efficient by re-accelerating the beam by means of the r.f. system of electron storage ring, so that it can be used over and over again. Segall [13] proposes to use a 1000 Å electron beam with energy in the range 500–1000 MeV to generate light with power in the

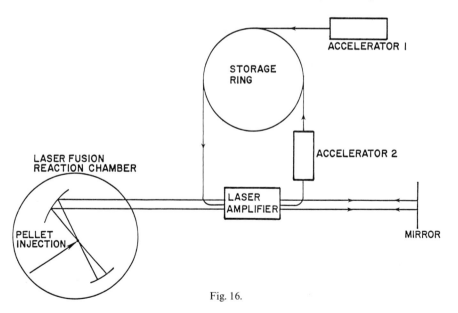

Fig. 16.

TW range. An efficiency of several percent is predicted. The scheme is illustrated by Fig. 16 which shows a storage ring from which electron pulses are passed into a laser system consisting of mirrors in the reaction chamber which focus the radiation, and a light path through the amplifier section leading to a reflecting mirror. When the laser energy in the focus has built up to the requisite intensity after successive beam passes, the pellet is propelled into it. Many modifications of this scheme are possible. Since the device operates *in vacuo*, it promises reliability, namely there would be no distortion or damage in transparent optical elements. It would be possible to tune the device to the optimum wave length, e.g. to a fraction of a micrometre. It is too early to say whether the scheme is feasible.

EPILOGUE

In this book, the physics underlying laser fusion devices has been explored in some depth. It forms the background of present and future research. Another monograph would be needed to discuss diagnostic instruments and laser system technology. The last chapter shows that great progress has been achieved in a comparatively short time. Large laser systems at present in the planning stage will almost certainly provide conclusive evidence concerning feasibility of controlled thermonuclear fusion. For a useful reactor, high efficiency of the laser system is crucial as we have seen in Chapter 2. The CO_2 laser is reasonably efficient, but the wavelength is long, and it was thought for some time that, with a critical surface so far away from the target, the coupling would be poor. Profile steepening might remedy this shortcoming but we clearly do not know enough about the energy transport. The search for a novel high efficiency and high power laser system at a wave length of one μm or less is of great importance.

REFERENCES

1. Giovanielli, D. V., Kephart, J. F. and Williams, A. H. (1976). *J. Appl. Phys.* **47**, 2907.
2. Giovanielli, D. N. (1976). *Bull. Amer. Phys. Soc.* Ser. II, **9**, 1047.
3. Rumsby, P. T. (1977). Rutherford Laboratory, private communication.
4. Lindmann, L. L. (1977). *J. de Phys. Supplement, Fasc.*, **12**, C6.
5. Ahlstrom (1977). Oxford Conference and private communication.
6. Tan, T. H., Giovanielli, D. N., McCall, G. H. and Wiliams, A. H. (1976). Proc. IRE Conference, Texas, IEEE, **95**.
7. Abel, N. H. (1881). Oevres, Paris.
   see also Whittaker, E. T. and Watson, G. N. (1948). "A Course of Modern Analysis", p. 229. Cambridge University Press, Cambridge.
8. Key, M. H., Eidman, K., Dorn, C. and Sigel, R. (1974). *Appl. Phys. Lett.*, 335.
9. Balmer, J. E. and Donaldson, T. P. (1977). *Phys. Rev. Lett.* **39**, 1084–1087.

10. Phillion, D. W., Lerche, R. A., Rupert, V. C., Haas, R. A. and Boyle, M. J. (1977). *Phys. Fluids*, **20**, 1892.
11. Perkins, R. B. (1978). 7th International Conference on Plasma Physics and Controlled Nuclear Fusion Research, Innsbruck, IAEA, Vienna.
12. Attwood, D. T., Sweeney, J. M., Auerbach, J. M. and Lee, P. H. (1978). *Phys. Rev. Lett.*, **40**, 184.
13. Segall, S. B. (1978). KMS-Fusion Inc., Report U806.
14. Motz, H. (1979). *Phys. Lett.* **69** A, in press.
15. J. M. Madey, D. A. G. Deacon, L. R. Elias, G. R. Ramian, H. A. Schwettman and T. L. Smith (1977). *Phys. Rev. Lett.*, **38**, 892.

INDEX